P9-BHX-981

Nutrition, diet and health

Nutrition, diet and health

MICHAEL J. GIBNEY
Senior Lecturer in Nutrition,
Trinity College, Dublin

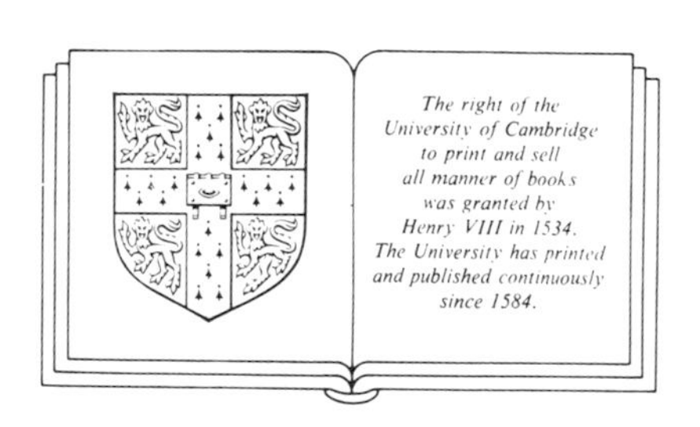

CAMBRIDGE UNIVERSITY PRESS
Cambridge
London New York New Rochelle
Melbourne Sydney

Published by the Press Syndicate of the University of Cambridge
The Pitt Building, Trumpington Street, Cambridge CB2 1RP
32 East 57th Street, New York, NY 10022, USA
10 Stamford Road, Oakleigh, Melbourne 3166, Australia

© Cambridge University Press 1986

First published 1986

Printed in Great Britain by
Billing & Sons Ltd, Worcester

British Library cataloguing in publication data

Gibney, Michael J.
Nutrition, diet and health.
1. Nutrition 2. Health
I. Title
613.2 RA784

Library of Congress cataloguing-in-publication data

Gibney, Michael J.
Nutrition, diet, and health.
Includes index.
1. Nutrition – Popular works. 2. Diet. 3. Health.
4. Diet in disease. I. Title. [DNLM: 1. Diet – popular works.
2. Nutrition – popular works. QU 145 G447n]

ISBN 0 521 30134 3 hard covers
ISBN 0 521 31756 8 paperback

RA
784
.G48
1986 / 50,918

SE

Contents

CAMROSE LUTHERAN COLLEGE LIBRARY

To Jo

Preface

Writing this book has been both fun and hard work. It has been prompted by a series of articles on diet and health which have appeared regularly in the 'Futures' pages of *The Guardian*. It is Mr Tim Radford, Features Editor of *The Guardian*, to whom I am most indebted for having encouraged me into writing this book. Professor Geoffrey Taylor, Emeritus Professor of Nutrition at the University of Southampton, has also played a major role in encouraging me to start writing on nutrition for the layman and has cast his hawkish eye over the typescript of this book. To both Geoffrey Taylor and Tim Radford I am grateful for their patient and friendly encouragement. Margaret Milner and Catherine O'Byrne at the TCD Medical School typed the original drafts and to them my thanks are due as they are to Dr Kate Younger for her meticulous criticism of the final draft. But it is to my wife Jo, to whom this book is dedicated, and who typed the final typescript, to whom I am most grateful, She has read endless drafts and showed infinite understanding and patience throughout.

Michael J. Gibney
Trinity College, Dublin
March 1986

1
Nutrition – a controversial topic: Hippocrates to hearsay

Food is not simply a physiological fuel. It is a social phenomenon. In celebrating we feast, in repenting we fast. We punctuate our calendar with festive foods, with turkey at Christmas and Thanksgiving, with eggs at Easter, with pancakes on Shrove Tuesday, and with pumpkins at Hallowe'en. In between we eat on average three meals a day and are subjected to a great deal of pressure on what we should eat, from manufacturers and advertisers, from experts and quacks, and from custom and habit. Nutrition is a topic of conversation on which the average person can hold an opinion, much as he holds opinions on nuclear weapons, the environment or taxation. Yet nutrition is unique among such controversies because the attendant opinion can be acted on at the individual level. A resident of Europe or North America cannot escape the threat of nuclear warfare, environmental decay and certainly not taxation. But they can change their diet any way they please without seeking permission, either by making an informed decision based on a sound understanding of nutrition, or by whim. Regrettably, these whims and the average person's misconceptions and misunderstanding about diet and health are played upon by an expansive array of nutritional advice which ranges from the supposed

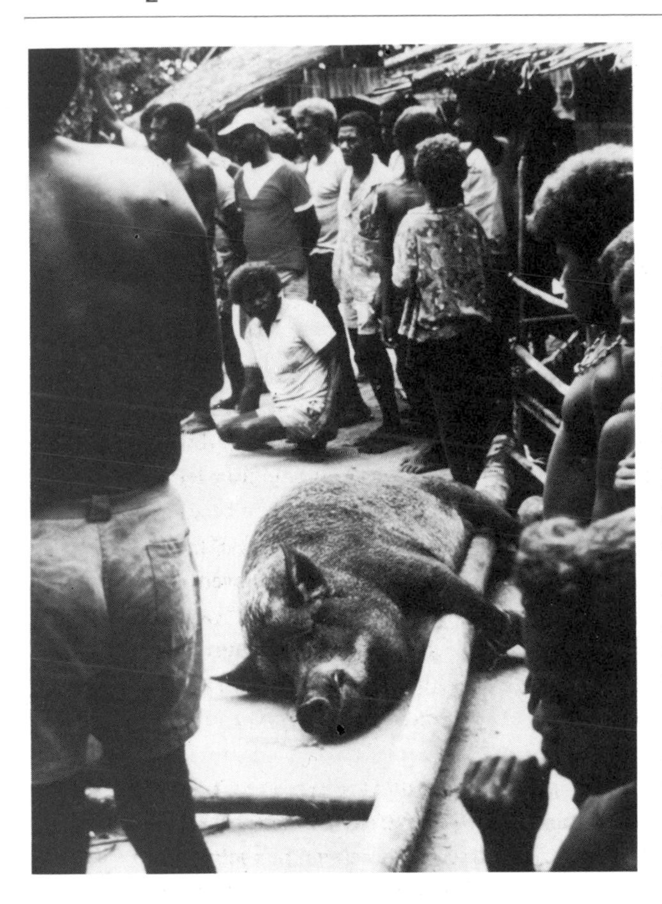

Fortunately, food has a long-standing and world-wide association with celebrations. These special occasions, such as the Solomon Islanders' Lambi Bay Feast, provide a welcome excuse to feast and indulge. Yet there is the danger that the public at large is being brainwashed by a plethora of gloomy missives into believing that food is a necessary evil, instead of a vital and enjoyable component of our daily life. (Reproduced with permission: COMPIX – Commonwealth Institute.)

expert to the downright charlatan. It is to help the average person to make sense of these conflicting views that this book has been written. It differs from most others on this topic in not offering any promise of a long or disease-free life. It sets out to explain the facts behind the controversies in nutrition, and no more.

Nutrition controversies are not new. The Greek philosophers linked the four elements of the cosmos – air, fire, water and earth – with the four humours of man – blood, bile, phlegm and black bile. Individual foods were assigned to humours. Accordingly, Hippocrates recommended pepper, combined with honey and vinegar, for the treatment of feminine disorders! The ancient Chinese also held diet high in their medical kit-bag:

'Experts at curing disease are inferior to those who warn against disease. Experts in the use of medicines are inferior to those who recommend a proper diet.' Whether Pan Ku, to whom these words of wisdom are attributed, knew then, or would know now, just what constituted the proper diet is debatable. The Egyptians believed the gut to be the root of all illness and sought considerable relief from the use of figs. The views of the Greek philosophers were to be taken up by the Salerno Medical School in Italy, run by the Benedictine Monks of Monte Cassino. Salerno exerted considerable influence in matters of medicine from the eleventh to the fifteenth centuries, probably to the detriment of its patients.

Such early views on diet were theoretical, arrived at by careful, logical thought. You can be sure that Hippocrates never conducted a double-blind trial on the efficacy of pepper, honey and vinegar for 'feminine disorders'! The first experimental evidence linking diet and disease was the relation between the incidence of scurvy aboard sailing ships and the lack of fresh fruit and vegetables. In the ensuing centuries, scientists employed the present-day approach of experimentation to understand digestion, salivation, respiration metabolism and the biological need for certain nutrients such as minerals, vitamins, amino acids and essential fatty acids. This quest extended well into the present century. Nutrition is a relatively new field of study.

Given the scientific nature of present-day nutritional research, one might wonder at the number and variety of controversies that still rage: the media regularly report on the conflicting advice concerning too much fat or salt or cholesterol, or too little fibre or exercise, or the dangers of additives and the benefits of multivitamin supplements. The average person is left baffled and sceptical about this conflicting advice. This is not the fault of science. It is because nutrition is good business, good copy for journalists, and good for the careers of some scientists and clinicians.

There was a time, just after the Second World War, when the Western world held science in awe. Science was the provider of wonder chemicals to anaesthetize, sedate, stimulate, cure diseases, grow plants, kill weeds, preserve food and prevent conception. Science was also the provider of the electronic revolution, from the television and telephone to the wonders of jet travel and computers. This confidence in science wained as science floundered in its own creations. Rachel Carson looked at the grim effects of pesticides and other pollutants on the environment, and young people queried the morality of scientific research to produce better napalm. The hippy movement epitomized a cultural revolution which sought a different lifestyle, independent of the triangle of state, military and industrial interests. At about this time the gurus came from the East, speaking in parables about peace and body and soul. The epidemiologists uttered their disgruntled noises about lifestyle and disease. The hucksters sneaked in to build their empires of deceit and the wagon trail of reform lumbered forward.

The interactions and effects of our diet and lifestyles on our health has proved to be an exceptionally tricky and complex problem. Many great scientists devoted their entire careers to this single and enormously difficult topic, and yet retained their scientific objectivity. Others were less fussy about evidence in their quest for easy solutions: they campaigned, and still do, for nutritional reform and governmental action; they became, and still are, celebrities of the media.

A former President of the Nutrition Society, Professor Ian Macdonald, wrote, 'Another problem associated with nutrition experts who are prepared to appear on television, is that quite often they may not be top-drawer experts. The reason for this is that interviewers on radio or television do not want statements with qualifications because they are not interesting and do not make news.' The pressure of these crusaders for nutritional reform, coupled with the sensitivity and confusion of the public on matters of health, created over forty expert committees in almost as many countries, who sat to peruse the evidence and issue their verdicts. In so doing, they were not acting as scientists but as public health officials and, as such, they operated by different rules. Scientists always keep one foot in the laboratory. Their philosophy is unique among those who quest for knowledge and, whilst Bacon, Popper, Medawar or Kuhn have written tomes on this philosophy, it is in John Le Carré's *Murder of Quality* that one finds their outlook best expressed: 'It had been one of Smiley's cardinal principals of research . . . not to proceed beyond the evidence. A fact, once logically arrived at, should not be extended beyond its natural significance. Accordingly, he did not speculate on the remarkable discovery he had made.'

Those scientists who have walked out of their laboratories and clinics to sit on committees and delineate public health policy did not have Smiley's qualms. They are, by and large, genuinely concerned and acting in good faith for all our interests. The net effect of their enthusiasm, the scepticism of laboratory scientists, the expedience of the media and the concern of the public has been to create the problem of present-day nutrition controversies. Other books may offer you simplistic answers. This one does not.

2
Nutrients and their metabolism

It is impossible to begin to appreciate the role of nutrition in health and disease without first considering the basics of what happens to the food we eat – its digestion, absorption, metabolism and excretion. This chapter tackles some of these aspects for the big three, proteins, carbohydrates and fats. You could skip this chapter, but your understanding would be hollow: how can you talk of increasing polyunsaturated fatty acid intake if you don't know what polyunsaturates are, or to try to understand protein deficiency in malnourished children if you know nothing of what makes one protein better than another or what makes a daily protein intake so vital. So stick with it.

Protein

Of all the major nutrients, proteins seem to have achieved a privileged place, being now, not least because of the advertisers, firmly associated with beauty, virility and cleanliness. This attitude is not without some justification. Unlike fat and carbohydrate, protein is an essential component of the diet. To go without is to invite trouble and to persist in this is to

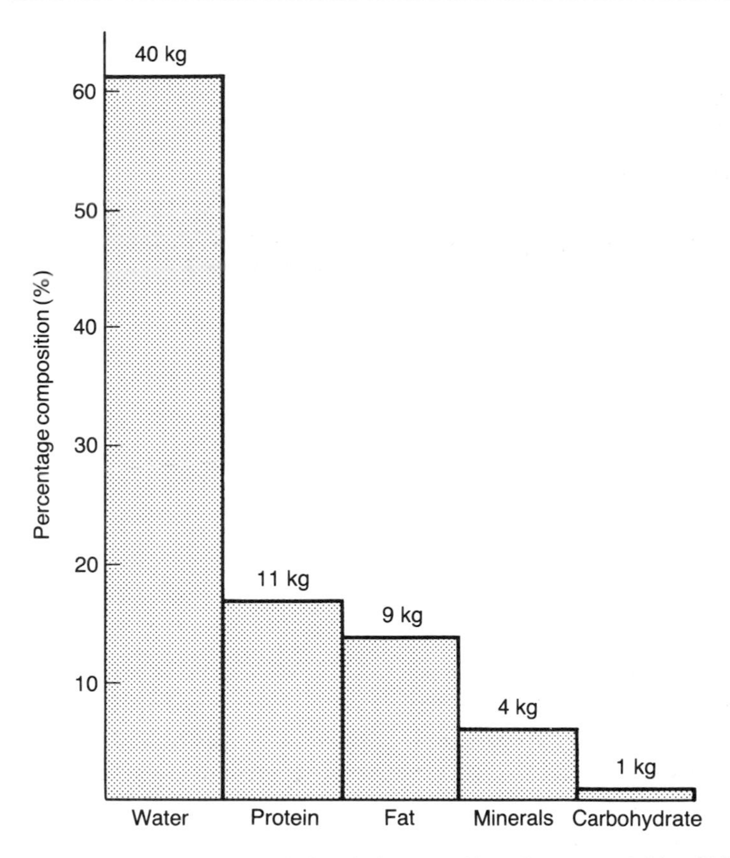

A bar chart showing the normal chemical composition of a man weighing 65 kg. The metabolism of the protein, fat and carbohydrate components of our diet forms the subject of this chapter.

court that form of death experienced by too many on this planet. It is the purpose of this chapter to discuss the nature of proteins, their function, their essential nature and the concept of protein quality. The problem of protein malnutrition will be dealt with later, in the proper context.

Proteins are the largest and most complex of the molecules in the body. They are constructed of several intercoiling strands. These strands are chain-like, the links being made of amino acids. There are 20 commonly occurring amino acids in nature and there is almost an infinite number of protein structures possible, for, in joining together amino acids to make the

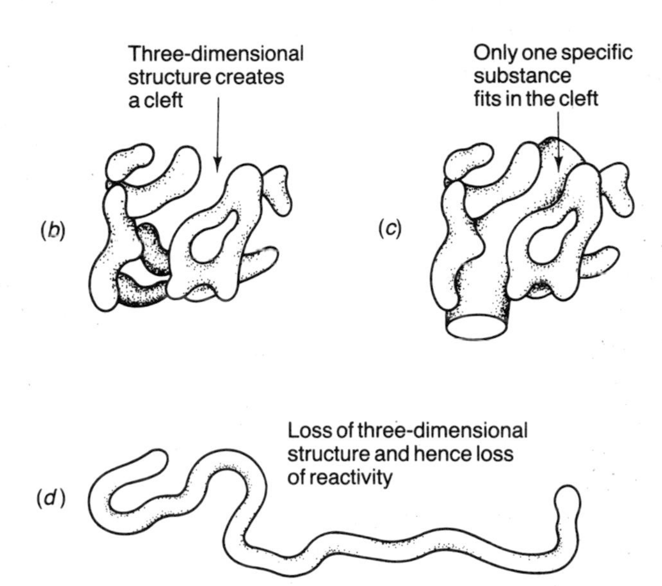

Proteins are made of chains of amino acids, as in (*a*). Long chains of amino acids become folded and cross-linked to give each protein its own distinct three-dimensional structure (*b*). In the case of enzymes or receptors, this permits the protein to react with a *specific* substance (*c*) for a specific purpose, e.g. to build, to break, to change. When a protein is 'denatured', such as by heat, heavy metals or acid, it loses its three-dimensional shape (*d*) and hence its reactive capacity.

component strands of protein, there are no ground rules. No one amino acid has to be present in any quantity and any amino acid can precede or follow another amino acid in the chain. Given that an average protein can contain 900 amino acids, it is obvious that a multitude of designs are possible. This versatility is biologically essential.

This is because almost every single molecular event in the body is regulated by an enzyme and because all enzymes are proteins. Given the staggering number of molecular events which occur in any cell at any one time, it is clear that a massive array of enzymes is essential. Hence the need for a variety of structures. But providing the body with enzymes is not the only function of proteins in the body and I shall mention just a few other examples. Quantitatively, the most abundant proteins in the body are those associated with the fabric of such tissues as muscles, skin, veins, arteries and tendons. Antibodies which defend against disease and the penetration of foreign elements are proteins, as are the particles which carry fat and cholesterol about the body. Blood clotting is dependent on a host of proteins as is the transport of the gases carbon dioxide and oxygen. What is common to all of these proteins is that each has one function and one function only.

It might seem strange that by juggling about with the order of amino acids in a protein chain, it is possible to effect such a variety of function and property. Perhaps an example would help. Gelatine is a protein. It is a crystalline structure which dissolves readily in hot water; when the solution is cooled, it forms a gel. The process can be reversed by heating. Egg white is also a solution of protein (ovalbumin) in water, but when it is heated it coagulates irreversibly. That the rearrangement of amino acids in a protein should achieve such major changes in properties is not so surprising if we compare it with the extraordinary variety and meaning of words achieved with minor rearrangements of just twenty-six letters: RIVET, RIVER, DRIVER, DRIVEL, SHRIVEL, SHIVER, SHOVEL, etc.

Proteins, therefore, differ from carbohydrates and fats in their variety of structure and their vital contribution to cellular life. There is another very important feature of proteins which gives them a special place in nutrition. One can quite happily exclude carbohydrates from the diet with no observable ill effects to most people under otherwise 'normal' conditions. Fats aren't quite as indispensable, although one could exclude them from the diet for quite a while without much ado. However, unless our daily protein requirement is met, we begin to deteriorate rapidly. Now it would not be unreasonable to ask why an average healthy person, endowed with all the right proteins in the right proportions, could not be allowed to pursue a protein-free diet in peace. Why, for example, does such a person, replete in all his proteins, need more protein, and daily at that?

The answer is that proteins do not last forever. Each protein has a definite lifespan, some lasting minutes, others months. An enzyme in the liver may last minutes, while a protein in the achilles tendon may last many

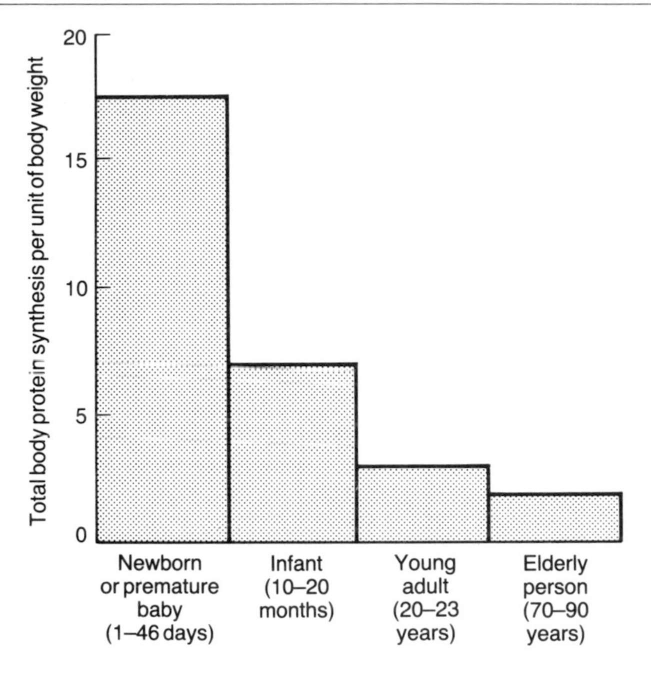

Proteins do not last forever. There is a continuous turnover of proteins in the body as old proteins are broken down and new ones are produced to replace those that have been lost from the body. The rate at which proteins are produced varies from one part of the body to another and also, as this chart shows, depends on age. A young baby has to synthesise a lot of protein to support its rapid early growth. A lack of protein at this stage in life can have serious consequences.

months. Of course, cells are also subject to turnover, so that, again after varying lifespans, they 'die' and liberate many proteins for recycling. All of this is known as protein turnover; it is central to protein metabolism and deserves some consideration.

If a protein is being synthesized and broken down continually, say in the liver, it is very easy to change its concentration rapidly: turn up the tap and close down the outlet and the 'tub' will rapidly fill (or, or course, vice versa). Thus, the first advantage of protein turnover is that as the requirements for a specific protein vary – and requirements vary rhythmically throughout the day – its concentration can be adjusted appropriately. It also makes sense that no one tissue should have a

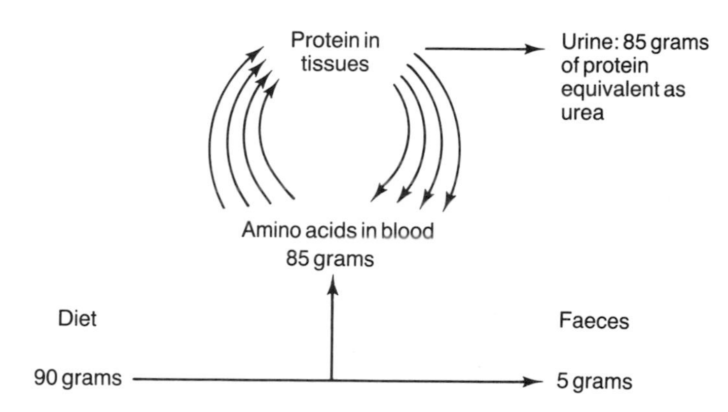

Protein equilibrium in a healthy adult, where intake equals output. The amount of protein synthesised and degraded is equivalent to about four times the quantity absorbed. This turnover continues even when protein intake falls critically. It is inherently inefficient and so urinary output of protein breakdown products always continues. Hence the need for protein intake on a regular basis.

monopoly on protein and this becomes very obvious on low-protein diets where muscle, a relative luxury, is broken down to provide amino acids for more vital proteins such as digestive enzymes and antibodies. Thus, protein turnover is to our advantage but there is a price we pay for this mobility and flexibility in protein redistribution. That price arises from the fact that the turnover of protein, like most things biological, is not 100% efficient; this leads to wastage and the excretion of protein breakdown products in urine and faeces. Clearly, since protein turnover happens whether we like it or not, and given the inefficiency of protein turnover, the excretion of protein breakdown products is something over which we have no control. So this excretion represents an obligatory loss of protein from the body, not as protein but as its major breakdown product, urea, in urine. It is because of this obligatory daily loss of protein that we have a daily need for intake of a comparable quantity. Without such an intake, the relentless turning over of proteins would leave us perilously deficient in these very vital molecules.

There is, in fact, a philosophical angle to the turnover of protein and other molecules of the body, even the constituents of the skeleton. They all display this phenomenon of turnover, inefficiency of metabolism and, hence, obligatory excretion. This means that the molecules which make up this year's body are not the same molecules that made up last year's model. The genetic code, which is passed on as cells divide in their turnover, ensures that this year's model looks like last year's, except of course one year older. It may look the same but it isn't. The molecules which made up last year's lush green pastures, its fruit and vegetables, its acres of cereals and pulses, are this year's Ronald Reagan, Mikhail Gorbachev, and you and I. Of course the route of molecular exit from the human sphere of molecules is much less interesting but should merit the occasional philosophical reflection next time you sit on the loo. The idea that the molecules which make up each one of us are in harmony with molecules in the plant world is intriguing. But enough of this philosophical distraction.

The proteins in the body are synthesized in all tissues, but some tissues are more active than others in this respect. The liver and muscles are probably the two most important protein synthesizers, on a quantitative basis. The brain and bone come low on this list. Indeed, the average amount of protein synthesized per person per day is 300 grams and the average intake is about 70 grams. This clearly indicates the dynamics of protein turnover, for the only way for the above to be true is for the body to make use of all the dietary intake at least four times each day. So a lot of synthesis and breakdown must go on.

The synthesis of a protein is a very finely controlled process, and is intimately associated with each individual's genetic code. Haemoglobin, the protein in the red blood cells which carries oxygen from the lungs to the tissues, is the same in all human individuals but haemoglobins vary in different animal species. This difference in protein structure between species has important implications for immunology and allergy which will be discussed later. The genetic code provides the blue-print from which the protein-synthesizing mechanism operates. It provides the information which causes the amino acids to be strung along the chain in precisely the correct order to produce the protein with the exact properties required. The next question is, where do these amino acids, which are the building blocks of proteins, come from?

They come from two sources, one inside and one outside the body. The latter is the dietary origin. The protein in our food is broken down by a series of specialized enzymes in the digestive tract. One group is secreted in

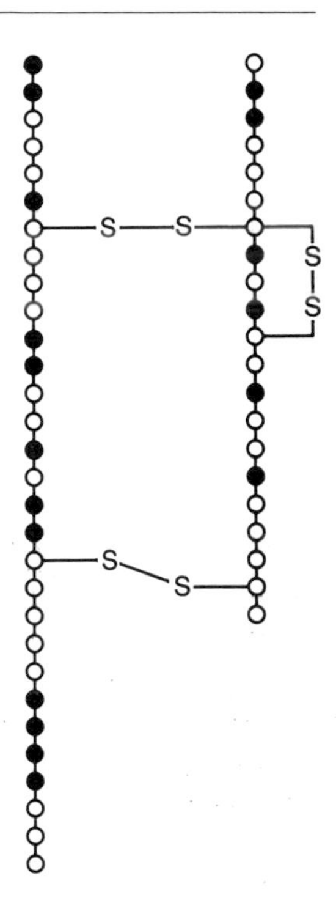

This illustration shows a protein molecule, in this case insulin, and the two different categories of amino acid of which it is made. The solid circles represent essential amino acids – these cannot be produced by the body and they must therefore be consumed in the diet for the body to be able to produce insulin and many other proteins vital to the maintenance of life. The open circles represent non-essential amino acids which the body is able to manufacture.

the stomach and the second group of enzymes for digesting proteins is produced in the small intestine. These enzymes cooperate, initially to fragment the large proteins of the diet, and then to reduce these fragments to even smaller units, until only the component amino acids remain. These are then absorbed across the gut wall and taken directly to the liver which acts as a GHQ for protein synthesis, taking its share of amino acids and then passing the rest into the bloodstream for the other tissues, such as muscles, to meet their requirements. The second source of amino acids is the body itself, which can make amino acids from appropriate by-products of carbohydrate metabolism. But it can only make ten of the twenty-two amino acids needed. These are known as the 'non-essential' amino acids (*see* Table 1). The remainder earn the title 'essential' amino acids because they are involved integrally in all proteins in the body, but the body *cannot* make them (Table 1). They must come from dietary protein.

Clearly, a dietary protein which is low in any of the essential amino acids

Table 1. *Essential and non-essential amino acids*

Essential amino acids	Non-essential amino acids
Isoleucine	Glycine
Leucine	Alanine
Lysine	Glutamic acid
Methionine	Histidine
Phenylalanine	Arginine
Threonine	Proline
Tryptophan	Ornithine
Valine	Aspartic acid
	Serine
	Tyrosine

is not much use to the process of protein synthesis, since the amino acids yielded on digestion will not permit production levels to reach their full target. If in the manufacture of cars there were a shortage of gearboxes, then there would be a proportionate shortage of cars. One option which might be open to the car manufacturer, but is not open to the body's protein-synthesizing mechanism, is that of building the car to a certain stage to await delivery of the gearbox. In the body it is whole proteins or none at all. There can be no compromise.

Some proteins are well endowed with the correct balance of the essential amino acids; others are not so well designed. Animal proteins, such as in eggs, meat and milk, show a good balance of essential amino acids. This is not surprising since these proteins have been made by animals quite similar, in the metabolic sense, to us. But many vegetable proteins show some amino acid to be in limited supply: in soya protein it is methionine; in cereals it is lysine. If a protein contains a limited amount of an amino acid, however, the disadvantage can be overcome by simply increasing the amount of protein consumed. This, alas, isn't always possible, as with young infants or in times of famine. To the average well-fed adult, vegetarian or otherwise, the 'problem' of protein quality is academic. To farmers producing fast-growing pigs and chickens, fed mainly on cereals and soya beans, protein quality is about financial profit. To those charged with the task of helping the Third World to veer away from the throes of famine, the problem of protein quality is vital and political.

Carbohydrates

Unlike proteins, the carbohydrates in the body contribute nothing to the structure of tissues and, although they contribute to the regulation of metabolism, they do not control individual molecular events as the enzymes (proteins) do. Their major function is the provision of energy to a variety of tissues, especially to the brain and nervous system which cannot utilize other nutrients for energy.

The carbohydrates in a typical breakfast – toast and tea with milk and sugar – are roughly representative of the distribution of carbohydrates in the average diet: starch from bread, potatoes, rice, pasta; sucrose from sugar; and lactose from milk. Starch is a large molecule made up of many glucose units joined together, all glucose units being of similar structure. It is rapidly digested to its basic glucose units which are readily absorbed. Lactose and sucrose are, by contrast, very much smaller molecules, each of which is digested to become effectively (in the liver) two glucose units. The enzymes responsible for their digestion are, respectively, lactase and sucrase. There is rarely a problem in the digestion of sucrose but a great number of people encounter problems with lactose digestion, most of which are associated with an inadequate supply of lactase. Undigested lactose passes from the small intestine, where digestion and absorption of its glucose units should occur, into the large intestine, where bacteria (a normal non-pathogenic population of microbes) ferment the lactose and cause digestive upsets and diarrhoea. The bulk of the population of Africa, southern Europe, the Near East, India and the Far East develop lactose intolerance during later childhood and adult life.

Under normal conditions, however, the great majority of carbohydrates in our typical meal are digested and absorbed as glucose. If you measured blood glucose levels before such a meal and at half-hourly intervals thereafter, you would see a rise in blood glucose, peaking at about the half-hour mark and returning to fasting levels almost as quickly. If you were to abstain from carbohydrates for a considerable period, say a week, your blood glucose levels would still be normal in spite of a minimal or zero intake. The body's capacity to maintain blood glucose within specific limits is achieved by a variety of hormones, the two most important of

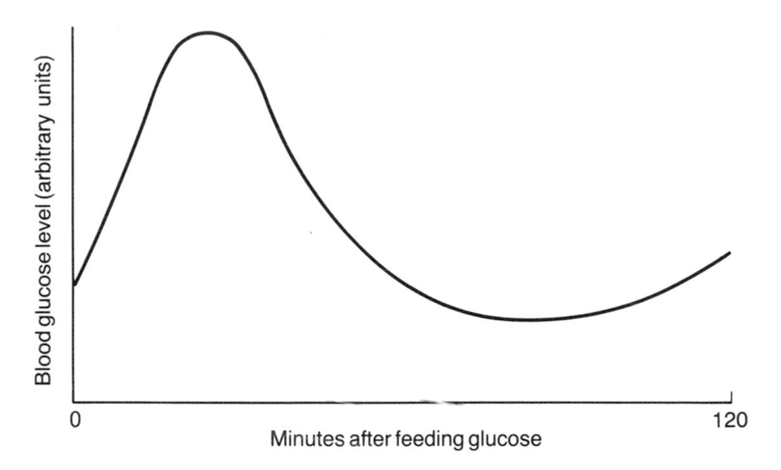

The level of blood glucose increases fairly rapidly after eating a meal but soon falls to near normal levels. The control of blood glucose levels is discussed further in the text.

which are insulin and glucagon. Both are secreted by the pancreas into the bloodstream, as required.

Insulin secretion is stimulated by a rise in blood glucose such as occurs after a carbohydrate meal. The main function of insulin is to remove glucose from the blood, the main recipients being adipose tissue (the body fat stores) and muscle. In the liver, glucose undergoes a molecular transformation from a carbohydrate to a fat, which is then stored around the body for use between meals. In muscle, the glucose units are reassembled into another carbohydrate, known as glycogen, which acts as a special short-term reservoir of energy for muscular activity. During all of this time, glucose is being transported into cells to undergo a step-by-step combustion to yield the energy for a host of biological and physiological functions. Suppose now that no more carbohydrate is forthcoming. Will blood glucose levels tumble as this tissue combustion of glucose proceeds? No. This is where glucagon becomes involved in maintaining normal blood glucose levels, stimulating the conversion of certain non-essential amino acids (glycine, alanine) into glucose, ensuring that none of this glucose gets stored as fat or glycogen, and forcing all tissues to switch from the combustion of carbohydrates for energy to the combustion of fats.

That, for the time being, is all we need consider for carbohydrate metabolism. Three aspects will be dealt with in greater detail in later sections: the role of muscle glycogen in exercise; the particular problems of diabetics in managing carbohydrates; and that ever-popular 'nutrient' – fibre.

Fats

If proteins have been dubbed 'Beauty' by advertisers, fats are surely 'The Beast', for few nutrients have been subjected to as much suspicion by the health profession. This suspicion is sometimes justified, although often for the wrong reasons. The confusion is understandable since the chemistry and metabolism of fats is more complex than that of proteins and carbohydrates, that is, up to the point that need interest us.

There are two obvious characteristics of fats about which we are all aware. First, they are not soluble in water, hence the need for detergents to clean up after a fatty mixed grill. Secondly, some fats are solid at room temperature, say butter, lard or margarine, while others are liquid, mainly the oils – olive, corn, peanut, etc. These are two good starting points for the study of the nutrition of fats, in that we must explain how fats, insoluble in water, are digested and metabolized in the largely watery environment of the body, and how the chemistry of fats can explain their physical properties.

Let us begin, therefore, with the chemistry of fats. In previous sections, it was said that amino acids are the backbone of proteins and that glucose is the backbone of the main dietary carbohydrate, starch. Later, we shall see that dietary fibre is also composed of large molecules made up of repeating smaller untis. The basic units of fat are known as fatty acids, but fat molecules are not made up of repeating fatty acid units. Fat molecules are known as triglycerides and are comprised of glycerol with fatty acids attached to the top, middle and bottom of the glycerol molecule. Hence fatty acids come in threes. Like proteins, there are no rules as to what fatty acids must occupy what positions on the glycerol molecule, or what mixes of three fatty acids should cling to a molecule of glycerol.

There are many types of fatty acids, all with a roughly similar, chain-like structure. However, for all intents and purposes, we need only consider a

(*a*) Stearic acid

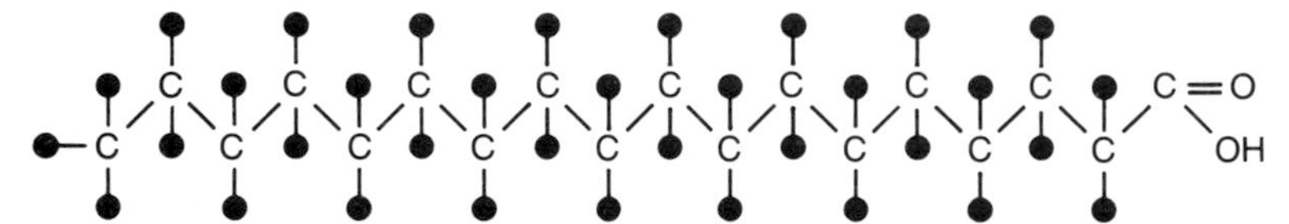

(*b*) Linoleic acid

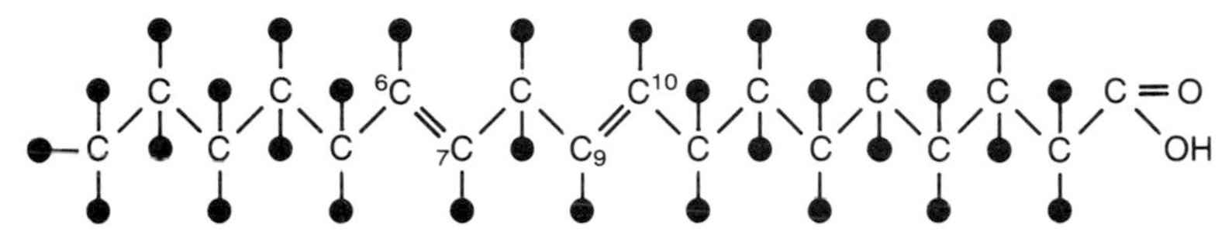

The difference in chemical structure between a saturated fatty acid (*a*) and a polyunsaturated fatty acid (*b*) might look insignificant but it does have important consequences for nutrition and health. This is discussed further in the text and also in Chapter 9 on Coronary Heart Disease.

(*a*) Stearic acid has 18 carbons, 36 hydrogens and 2 oxygens in its structure. Each carbon atom is 'saturated' in that its requirement for four complete bonds is met.

(*b*) Linoleic acid also has 18 carbons and 2 oxygens in its structure but has only 32 hydrogen atoms. Thus the carbons at 6 and 7 and at 9 and 10 are missing a hydrogen, necessitating a double bond (C=C) between adjacent carbons. Such bonds are 'unsaturated' (in hydrogen atoms). With two or more such bonds the fatty acid is deemed polyunsaturated.

dozen or so and, for the time being, we need only consider three groupings of these fatty acids: saturated, monounsaturated and polyunsaturated fatty acids.* The terms saturated fats and polyunsaturated fats are commonplace but are used imprecisely. No dietary fat source is comprised solely of fatty acids of one category. For example, butter is regarded as a saturated fat but is in fact comprised of 65% saturated, 32% monounsaturated and 3% polyunsaturated fatty acids. Margarines are notoriously variable in composition but a typical soft margarine is comprised of the following fatty acids: 31% saturated, 47% monounsaturated and 16% polyunsaturated. So, lesson number one is that the terms saturated or polyunsaturat-

* A fatty acid is 'saturated' if all the possible sites for hydrogen atoms are occupied. If a fatty acid molecule has a vacancy for one pair of hydrogen atoms, it is 'monounsaturated'; if there is room for more than one pair of hydrogen atoms, the molecule is 'polyunsaturated'. In the manufacture of margarines, hydrogen is added to a polyunsaturated oil; the final composition depends on the amount of hydrogen added.

ed are useful but somewhat misleading, as are the terms animal and vegetable fats.

If you were to purchase samples of pure saturated fatty acids from a chemical supplier you would find that they were solid at room temperature. Monounsaturated and polyunsaturated fatty acids would be liquid. Therefore the hardness or softness of a dietary fat depends on temperature and the balance between the three categories of fatty acids. Now let us consider fat digestion.

The enzyme responsible for fat digestion is called lipase and is secreted into the small intestine. The problem is that lipase is found in the watery aspect of the gut while fats, being incompatible with water, either float about on top or travel about as small droplets, as in, say, french dressing.

Therefore, in order for fats to be digested by lipase, they must become finely dispersed in the watery gut contents, much as oil spillages in the sea must be dispersed so that we can gain access to them and break them down. The dispersing agent in the gut is bile, which is made in the liver and secreted from the gall bladder. The bile salts are powerful detergents and permit each fat molecule to be broken down by lipase to its basic components, i.e. glycerol plus three fatty acid molecules. These are then transported into the gut wall where they are re-formed, though rarely in the precise arrangement in which they started. Now there is another problem. How do you transport fat molecules through what is basically a watery bloodstream? To do this, the body coats a tiny cluster of fat molecules with protein and phospholipids, a special class of fats. Phospholipids enjoy 'dual nationality', i.e. one end of the molecule preferably sinks into the fatty core, while the other end prefers to protrude into the watery environment.

Thus are fats transported in the body. The composite particle of fat droplet + protein + phospholipid is known as a lipoprotein and these lipoproteins are, as we shall soon see, at the very centre of heart disease. The lipoprotein is carried to the fatty deposits of the body – the flab.

In the previous section on carbohydrates I said that carbohydrates give rise to fats, i.e. they are fattening. This transformation of carbohydrate takes place in the liver and presents the liver with the same problem as the gut, i.e. how to transport the newly synthesized fat about the bloodstream; the remedy is identical. A lipoprotein particle is secreted with the cluster of fat droplets in the centre, surrounded by a thin shell of protein and the 'dual-national' phospholipid and again this lipoprotein deposits its contents into fatty tissue. So if the liver can make fat from carbohydrate, is there any need for dietary fat? Is this ugly nasty necessary and if not could it be taxed or labelled by the government with health warnings as with cigarettes? Alas not! For the liver, talented as it might be, can only synthesize the saturated or monounsaturated fatty acids. It cannot make polyunsaturates. And polyunsaturated fatty acids are essential for life.

Elsewhere in this chapter, I have deferred several detailed discussions to later sections. Just this once let us indulge in a short interlude of more detailed discussion of this phenomenon.

Essential fats and prostaglandins

Given that polyunsaturated fatty acids are essential to life and given the inability of the liver to make them, there is only one source – the diet.

The essential nature of polyunsaturated fats is an interesting story. Back in the 1930s, nutritionists were obsessed with identifying exactly what components of the diet are not dispensible. The essential amino acids were identified. The vitamins, minerals and trace elements were unearthed one by one. Two Californian scientists, George and Mildred Burr, found that rats fed diets from which fat was excluded failed to grow properly and developed unusual skin lesions. Being the methodical duo they were, George and Mildred carefully added, in turn, all the other then-known essential nutrients back into the rats' diet. Even then there was no improvement. Clearly there was something else, something in fats, that was somehow essential for normal growth and development. So with due ceremony, in the attic of a Berkeley laboratory, they dubbed this unknown compound 'vitamin F'.

Careful investigation then showed that vitamin F was in fact a polyunsaturated fatty acid – linoleic acid.* It is in fact the most abundant polyunsaturated fatty acid in nature. Further research began to throw some light on its essential nature. The liver takes linoleic acid – the dietary essential fatty acid – and converts it to another fatty acid called arachidonic

* If you read your food labels carefully you will be familiar with this term. Have a close look at a tub of margarine, which makes the label claim 'polyunsaturated'.

acid, which is slightly longer and twice as polyunsaturated as the precursor, linoleic acid.

Where is arachidonic acid found in the body? 'In the fatty tissue like all other fatty acids that we absorb from the diet or manufacture in the liver', you might reply. Yes, that is so, but only in minute quantities. After its manufacture in the liver, most arachidonic acid is incorporated into cell membranes.

The basic unit of life is the cell. Some forms of life exist as single cells. Others are more complex, although they usually start with two simple cells (a sperm cell and an egg cell) and then develop into the complete plant or animal, made up of many organs and tissues, each interlinked and each comprised of many millions of cells serving specialized functions. The cells arc themselves compartmentalized, with each compartment doing its own thing. To retain each function distinct from the next, each compartment is bounded by a membrane, as is each cell and then each tissue, and so forth.

The membranes are composed of a mixture of compounds including proteins (once again we see the specialized functions of proteins) and our friendly 'dual-national' phospholipid. Each phospholipid has two fatty acids attached, and about 33% of these fatty acids are arachidonic acid.

It was another 40 years before the next phase of the story was revealed, that is, just what is so special about the arachidonic acid in the phospholipids of the multitudinous membranes in the body. Indeed the workers responsible for this enlightenment earned the 1982 Nobel Prize in medicine. Arachidonic acid is converted, when required, into a group of chemicals with a staggering role in regulating such essential processes as the maintenance of blood pressure, the contractions of childbirth, the clotting of blood, the motion of the gut, and the defence against foreign bodies. These compounds are known as prostaglandins. In absolutely minute quantities, and with a viability measured in seconds, they regulate a diversity of bodily functions. Because they are so potent and so short-lived, they are only made where and when required. Hence the need for a local supply of arachidonic acid from which they come. Simple isn't it? No fat means no polyunsaturated fatty acids, no linoleic acid, no arachidonic acid, no prostaglandins and no future!

The average daily requirement for the essential fatty acids is 10 grams per day for the average adult and that errs on the side of caution by several fold. It is not impossible, but it is ridiculously difficult, to fall below this level, given even the most spartan of normal diets, so please pick something else for your *Guinness Book of Records* ambitions.

Some people reading this book will now begin to think along the following lines: polyunsaturated fatty acids are essential for life; 10 grams a day is deemed the desired intake; therefore, 20 grams a day is twice as good and 30 grams a day is three times as good as just the basic 10 grams. That is the most common error in people's attitudes to diet. Once your needs are met, any excess is surfeit and unnecessary. You cannot improve on the quality of an egg by having three shells for one yolk.

And so ends this interlude which, hopefully, has whetted your appetite for more of the science behind nutrition.

Fibre, cholesterol, etcetera

This chapter has been concerned with the metabolism of fats, carbohydrates and proteins, for these are the only nutrients capable of yielding energy for the many essential functions of the body. But there are, besides these nutrients, a host of other components such as cholesterol, fibre, vitamins, minerals, trace elements, food additives and naturally occurring toxins. These will be discussed in later chapters.

Summary

PROTEINS are composed of intercoiling strands, in turn composed of amino acids.

- The proteins function to regulate specific steps in metabolism – one step, one protein. Hence, many proteins are needed.
- The variety of protein structures is made possible by the order and arrangement of the amino acid building blocks.
- At any time, body protein is being both synthesized and degraded, an inefficient process which leads to urinary loss of the breakdown products of the amino acid building blocks.
- This loss is obligatory and necessitates regular replenishment – hence the nutritional necessity for protein in the diet.

CARBOHYDRATES are composed of sugar molecules, single, paired or in long strands. Dietary carbohydrates are predominantly starch, with some sucrose and lactose.

- Starch is digested to glucose units, which are observed to be used for energy or for the synthesis of fats.
- Rising blood glucose levels after a meal stimulates insulin release which disposes of this load. Falling blood glucose levels stimulate glucagon, which maintains a basic minimum level of glucose in the blood.
- Although sucrose is well tolerated, lactose is poorly tolerated by many people who lack the enzyme to digest it. Both these substances yield glucose or glucose-like compounds on digestion.

FATS are comprised of fatty acids bound, in bundles of three, to glycerol.

- Fatty acids are either saturated, monounsaturated or polyunsaturated.
- Fats contain mixtures of all three types of fatty acids; the balance of the mix determines whether the fat is generally saturated (hard) or polyunsaturated (soft).
- Fats are insoluble in water and require special treatment for digestion and transport.
- In digestion, bile acids act as detergents to solubilize fats to aid digestion.
- In transport, the fatty material is surrounded by a coat of protein plus phospholipid. The full particle is a lipoprotein.
- The liver can convert carbohydrate to fat, which is exported from the liver as a lipoprotein.
- Certain polyunsaturated fatty acids are nutritionally essential because: (1) they are the precursors of the vitality important regulators of metabolism, the prostaglandins; and (2) the body cannot make them from either carbohydrate or existing dietary fatty acids.

3
Fibre

Fashion is not confined to the clothes industry, to holiday resorts or to political belief. There are fashions in most professions, including science. In general, a subject becomes fashionable when scientists see in it an opportunity to make a name for themselves by publishing papers and to secure research grants. This is easiest in a new field or, as in the case of fibre, an old field which has been revived.

Adam and Eve started the ball rolling by eating an apple. Fibre intake was low in the Stone Age when man hunted animals for food, but rose when early agriculture became established. For fibre is found only in plant tissue, and the more unprocessed the food, the more fibre is present. The ascent of Western man brought with it a degree of affluence such that people consumed less plant-derived food and more foods of animal origin. That has always been the case. Celebrations and feasts are characterized by meat (and booze!) and always have been. The poor, besides envying the rich man's food *per se*, envied his diet because of its exquisite refinement. By the time the twentieth century had arrived, most people in the West, in peace time at least, could enjoy the rich man's diet of meat, milk, eggs, cheese, nice white crustless cucumber (peeled of course) sandwiches, long-grain polished rice or apples tastefully peeled, cored and quartered. Meanwhile, people in the less prosperous regions of the globe ate as they always had – infrequently, and on a diet of raw, unrefined plant produce such as yams, cassava, rice, maize, millet, wheat or beans. Some eminent physicians working among these people were struck by the rarity of the very diseases which were most prevalent in Western society: high blood pressure, stroke, heart disease, tooth decay, obesity, bowel disorders and diabetes. Their initial observations were confirmed by scientific studies. A new catch-phrase arose, 'diseases of civilization', and smart scientists and clinicians soon realized that there was a future in fibre. Fibre had been rediscovered.

What was the conventional attitude to fibre then, in the early 1960s? In general it was to ignore fibre as being of no importance in medicine. Indeed, these early investigators would have been perplexed by the term then used to describe fibre, i.e. crude fibre. This phrase had been coined by animal nutritionists to help them to predict the food value of grassland. It is, therefore, not surprising that the clinicians questioned the wisdom of using this measure of fibre in human nutrition and set out to answer the question of what exactly fibre, or crude fibre, was, in botanical and chemical terms.

Plant food is basically divisible into stems, fruits, tubers, seeds and leaves. Stems are rich in fibre to support the upper leaves and connect them with the nutrients in soil. The leaves are green, rich in the protein chlorophyll which traps the sun's energy on which plants depend. The fibre content of leaves and stems varies, tending to increase with age. Seeds, fruits and tubers come in a variety of shapes, sizes and colours.

A seed has an outer coat to protect the embryo as the seed lies dormant during the winter. Come the warm days of spring and the seed begins to germinate, pushing a new shoot through the now cracked outer coat to reach the soil surface and point its tiny leaves towards the sun. In the period between the onset of germination and the emergence of the seedling into the sunlight, the energy reserves of the seed are used up in growth. Some seeds (e.g. soya) use oil for these energy reserves, but most (e.g. cereals, rice, cassava, potatoes) use starch. Although there is some fibre in these starchy stores, it is present in much greater quantity in the structural part of the seed – its protective coats and its plant embryo, or 'germ'. It is these coats and germs which man chooses to remove and discard, so that his

a

b

c

There has been a trend in the affluent countries of the developed world to produce and consume a diet that is progressively becoming more refined, more processed and further removed from staple plant produce.

In contrast, earlier civilisations, and the less-developed countries of the world today, have always relied on basic plant produce as the staple element in their diet: (*a*) this illustration from a tomb at Thebes shows the threshing, winnowing and porterage of grain some 3500 years ago by the Egyptians; (*b*) another important crop, the potato, is here shown being cultivated and collected in Inca times (from the Poma de Ayala MSS. reproduced by the Institute d'Ethnologie); (*c*) bread, whether leavened or unleavened, is still an important element of the diet in many countries of the world as shown in this baker's shop in Afghanistan with its sunken clay oven.

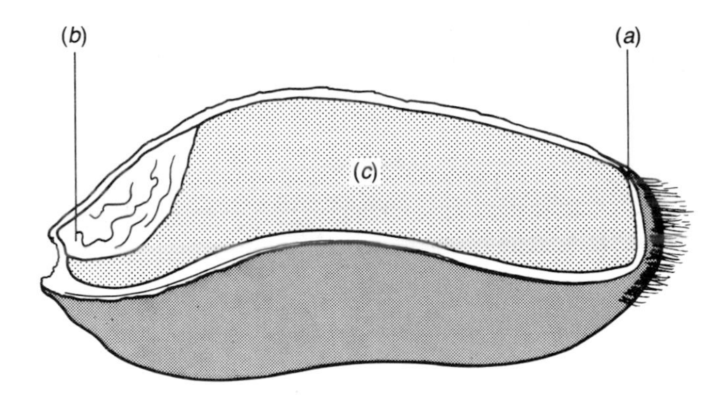

This wheat grain shows the three main components: (*a*) the outer coat is the structural part that is rich in fibre; (*b*) the germ or embryo plant is rich in protein, vitamins and minerals and also has some fibre; (*c*) the inner portion of the seed is a source of starch for the growth and development of the plant embryo.

White flour and white bread is made by carefully removing (*a*) and (*b*) from the grains and thus results in a low-fibre product. In contrast, wholemeal flour and bread utilise all of the grain and thus retain the natural fibre of the wheat.

enjoyment of the starch core is unimpaired. That is a botanical description of fibre; now for the chemical description.

There are four main molecular species which make up fibre: cellulose, hemicellulose, pectin and lignin. The first three can be classified as carbohydrates, lignin being unclassifiable in the nutritional sense. Cellulose, for example, is comprised of long chains of glucose molecules joined end to end. You will remember that was exactly how another carbohydrate, starch, was defined. However, the way in which the glucose units are joined together is different (see illustration on p. 28).

The difference is very small, but the enzymes that digest carbohydrates have lines of demarcation that put even the most militant shop steward in the shade. Amylase, the main carbohydrate-digesting enzyme in the gut, will only act on starch, never on cellulose. So cellulose is not digested. Neither are hemicellulose, pectin and lignin. And that is the biological definition of fibre in human nutrition: indigestible plant matter, particularly indigestible plant carbohydrates. Table 2 lists the distribution of fibre fractions in three common foods.

Clearly, what is fibre in one food is not altogether the same fibre in

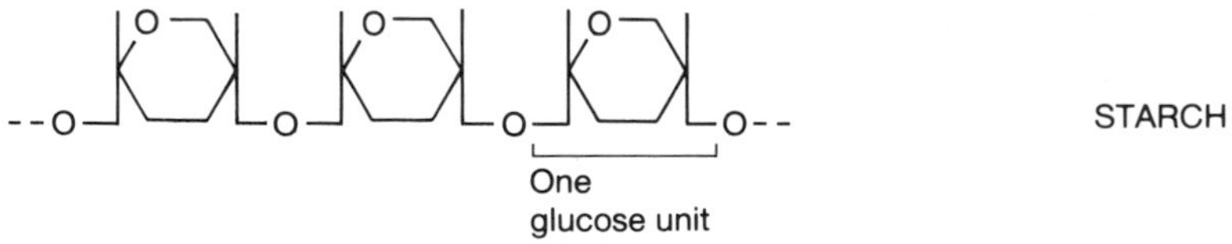

STARCH

One glucose unit

CELLULOSE

Cellulose and starch, though both composed of glucose subunits, have slightly different structures. Whereas starch is readily digested to produce glucose, cellulose is not digested at all and passes straight through the gut as undigested fibre.

another food. Since the physiological effects of these different fibre fractions vary, it follows that you cannot expect cabbage fibre to achieve the same result as bean fibre or apple fibre.

Let us return to the terminology of fibre. The term 'crude fibre' refers to that part of the plant carbohydrates which resist gentle boiling in acid and then in alkali, a process believed by the agricultural chemists of the last century to resemble digestion! The system works for cattle and sheep, for which it was designed, because they possess the capacity to digest all plant carbohydrates – starch, cellulose, hemicellulose and pectin.* A new system of chemical definition had to be established and, amid much controversy, the phrase 'dietary fibre' emerged to define the non-starch plant carbohydrates (to include lignin, but not lactose or sucrose, i.e. milk sugar and table sugar). The distribution of carbohydrates is outlined schematically in the illustration opposite. Nevertheless you may see, even today, on food labels and in nutrition reports, the term crude fibre where dietary fibre should be used. Beware the difference.

Now let us consider the physiological effects of fibre, beginning with the most obvious: the effect of fibre on stool frequency and consistency.

* This is only made possible by the bacteria which live in the first stomach of cattle and sheep and secrete enzymes to digest any and all carbohydrates except lignin or lignin-entrapped carbohydrates. Man has an equivalent bacterial population at the lower end of the digestive tract which will be discussed shortly.

Table 2. *Percentages of fibre fractions in three common foods*

	Cabbage	Apple	Wheat
Cellulose	63	33	19
Hemicellulose	trace	5	7
Pectin	26	36	65
Lignin	10	26	9

One of the most readily demonstrable properties of fibre is its capacity to absorb water and to swell: 1 gram of rhubarb fibre combines with 15 grams of water; and 1 gram of bran combines with 5 grams of water. This means that individuals on high-fibre diets not only have a greater stool bulk (because their intake of inherently indigestible material is high), but they also have a softer stool because of its higher water content. These softer and bulkier faeces are much easier for the body to propel along the large intestine and eliminate from the rectum. Faecal consistency is measured ingeniously. A 150-gram cone is suspended just above the stool, almost touching it. It is held there for 5 seconds and suddenly released. Penetration into the stool is measured, gingerly one suspects, in millimetres. The softer the stool, the greater the penetration. Intestinal transit times can be measured using a variety of markers; these are added to

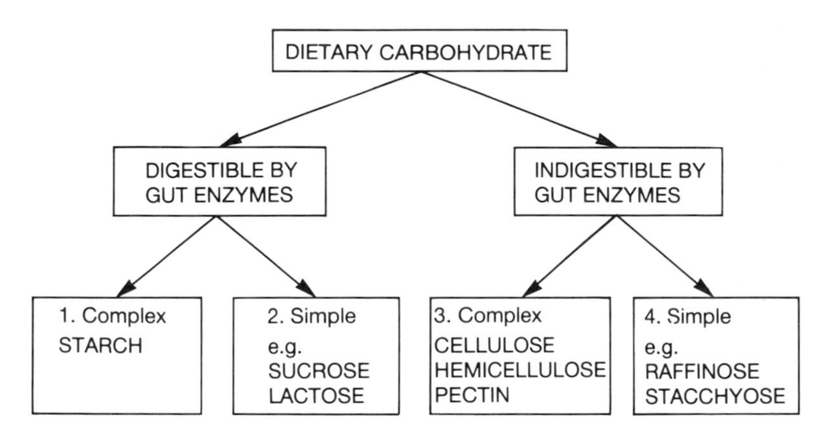

The distribution of dietary carbohydrates. Dietary fibre includes those in boxes 3 and 4 and also lignin.

Table 3. *Crude and dietary fibre intake*

	Average intake per capita per day (UK) (grams)	Wholemeal bread (grams per 100 grams)
Crude fibre	5	1.1
Dietary fibre	25	5.1

a test meal and their appearance in the faeces monitored. The average transit time of material through the gut is 96 hours, with a faecal weight of 75 grams per day. By increasing faecal weight to 100 grams per day, the transit time is reduced to 75 hours, and will fall as low as 36 hours with faecal weights of 200 grams per day. One of the effects of a decreased transit time is that internal gut pressure is reduced and, although it is not possible to establish a direct link between transit time or gut pressure and disorders of the gut, it is possible to provide a biological basis to the established statistical evidence linking low-fibre diets with appendicitis, diverticular disease and haemorrhoids (piles). For example, forty patients suffering from diverticular disease were each given three heaped tablespoons of bran per day on top of their normal food for a 6-month period. In 60% of patients, symptoms were abolished by this treatment, 10% showed no change and 7% were made worse. Diverticular disease affects 35% of the over-60s in Western countries and frequently leads to surgery, so its successful treatment with fibre is an important discovery. A second feature of decreased transit time is the possibility that the potential of naturally occurring or synthetic compounds to exert a carcinogenic (cancer-causing) effect, particularly in the large bowel, is reduced, although much work is still required to clarify this.

Let us continue to review the physiological effects of fibre in the lower regions of the gut, before moving upwards to the small intestine and stomach. The large intestine is inhabited by a colony of bacteria. It is perfectly normal for these bacteria to be there, and they probably contribute to our well-being. They live on whatever arrives down in their region of the gut, which inevitably means that part of our diet which isn't or can't be digested in the stomach or small intestine. Hence, fibre supplies a large part of their diet. In fact these bacteria can break down dietary fibre into simpler substances, some of which are absorbed and utilized by the liver and other tissues for energy. So, strictly speaking, some fractions of fibre (which by definition is indigestible by man) are digestible by microorganisms and some of that material can be used by man. However, what are more important to the fibre-eater are the by-products of this digestion. These bacteria live in an environment devoid of oxygen, not unlike the yeast lying on the bottom of a demi-john of fermenting wine. Just as wine- or beer-fermenting yeasts produce gas during the fermentation reaction, so do the bacteria in the large intestine. The amount of gas (carbon dioxide, methane and hydrogen) produced depends on the amount and nature of fibre delivered to them. They are especially receptive to those carbohydrates shown in category 4 of the previous illustration. These are simple molecules, readily soluble in water and therefore easily accessible to the bacterial enzymes which bring about fermentation. They are found in sizeable quantities in legumes (peas and beans). Hence the basis for the Dublin street-song:

> Beans, beans, good for your heart,
> The more you eat, the more your fart,
> The more you fart the better you feel,
> Beans, beans at every meal.

Breaking wind is one of the drawbacks of a high-fibre diet and has always been socially unacceptable. The Roman emperor Claudius is said to have planned an edict to legitimize the breaking of wind at table, either silently or noisily, after hearing of a man who was so modest that he endangered his health by an attempt to restrain himself. The Salerno diet, as advocated by the Crusaders, claimed that:

> Great harms have grown and maladies exceeding,
> By keeping in a little blast of wind:
> So cramps and dropsies, colics have their breeding,
> And mazed brains, for want of vent behind.

Besides social graces, there are times when breaking wind is downright dangerous, such as when in a spacesuit; some small part of the NASA Nutrition Research Programme may be concerned with this. Let us now move on and discuss the physiological effects of fibre in the stomach and small intestine.

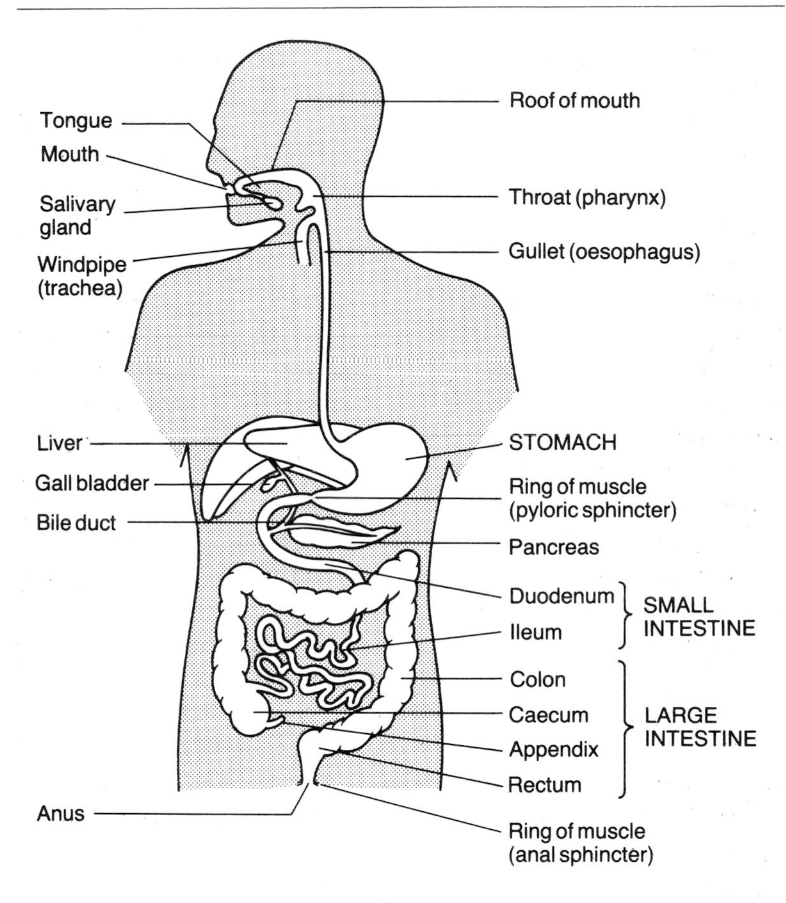

The human gut showing the major divisions into the stomach, small intestine and large intestine.

It is said to be bad manners to leave the table feeling full. Be that as it may, I doubt that social graces are the overriding factor in deciding when we've had enough. The regulation of food intake is a complex biological matter. The feeling that we've had enough to eat is known as satiety and it is regulated by hormones acting on a specific part of the brain (the satiety centre). The acts of chewing and swallowing, and the degree of distension of the stomach after a meal, together with the hormones released, are believed to stimulate the satiety centre in the brain to switch off feeding.

That is the theory. Clearly, the more chewing and swallowing involved, together with greater distension of the stomach, the more rapidly the satiety centre is activated and the sooner we feel full. Whether we choose to listen to our satiety centre is another matter. Common sense will tell us that, for example, croissants are 'less filling' than wholemeal bread, and experiments bear this out. In another experiment, volunteers consumed either 450 grams of apples, or the same amount of apples purĕed to a fine pulp, or apple juice containing the same amount of carbohydrate as 450 grams of apples. They then recorded how 'full' they felt, using a satiety score. The average score was +7 for the intact apples, +5 when chewing was rendered unnecessary by pulping, and −1 when the fibre-free juice was given. Other experiments have verified these findings, and lead us to conclude that high-fibre foods which require considerable chewing and which provide bulk in the stomach, satisfy our hunger more quickly than do refined foods. There is considerable sense in the theory of the 'F- plan' diet. Whether it will work for you will depend on your particular metabolic idiosyncrasies as outlined in Chapter 4. But that's no excuse for not trying.

There is little doubt that it is in the investigation of the dietary management of diabetes that fibre has made most impact on medicine. Diabetes, is caused by a deficiency of the hormone insulin which, as described in Chapter 2, clears glucose from the blood after carbohydrate (starch, sucrose or lactose but not dietary fibre) is eaten. The glucose is either directed into muscle, to supply energy needs, or stored as fat, for subsequent use by all tissues. There are two types of diabetes. Type I usually develops in adolescence and is characterized by an absolute deficiency of insulin, requiring daily injections of the hormone. Type II develops in adulthood and is due to only a partial failure in insulin production; it is therefore less severe, although it is much more frequent than type I. Furthermore, about 75% of type II diabetics are obese and so the treatment of this type of diabetes very often involves weight control. In both types of diabetes, fibre has been found to play a major role in the management of the disease. Normally this management involves the regulation of the type of carbohydrates consumed. Sugar is taboo because it is a highly concentrated form of carbohydrate, releasing glucose rapidly into the bloodstream. Starch intake is usually limited to no more than 40% of total daily calories, which compares with a generally recommended value for non-diabetics of 60% (although the actual value is nearer 50%). The resurgence of interest in fibre has turned this outlook on its head. It began with the observations of those same doctors who first noticed that many of the diseases of civilization were relatively scarce in well-fed communities of the Third World, whose diet provides 70% of total calories as carbohydrate. Furthermore, they noted that those who did develop diabetes did not seem much the worse for this very high carbohydrate intake. So they set about the task of looking at what would happen to diabetics in Western societies when placed on a high-fibre diet. In general, they found that diets rich in complex carbohydrates (categories 1 and 3) slowed down the release of glucose into the bloodstream, as shown in the illustration opposite.

The chemical basis of how fibre acts to slow down the release of glucose into the bloodstream is not at all clear. It may be that it physically impedes the passage of glucose across the gut, or perhaps the bulky fibrous material dilutes the access that glucose has to key hormone-activating sites along the gut. Either way it must be said, as always with fibre, that different sources (having different proportions of the four components of fibre: see Table 2) have different effects on the release of glucose into the bloodstream. The most successful of these is guar gum, obtained from the cluster bean. For years it has been used in the food industry as a thickening agent in, for example, ice-cream. There has recently been a hectic scramble to incorporate guar gum into everyday foods for diabetic use. However, the preferred method is to pursue a high-legume diet, which is what the British, American and Canadian Diabetic Associations now recommend for their members. Furthermore, these bodies believe that the previously imposed upper limit of 40% of calories from carbohydrate should now be waived. This is good news for diabetics, at least those who like high-fibre foods like legumes. It doesn't represent a cure, merely the improved management of the disease.

There are two other diseases, the two main contributors to mortality in Western society these days, which are linked with fibre intake: coronary heart disease and cancer. Studies have shown that those individuals with the highest intake of fibre have the lowest incidence of both these diseases, and vice versa. However, as we shall see later in a more detailed discussion

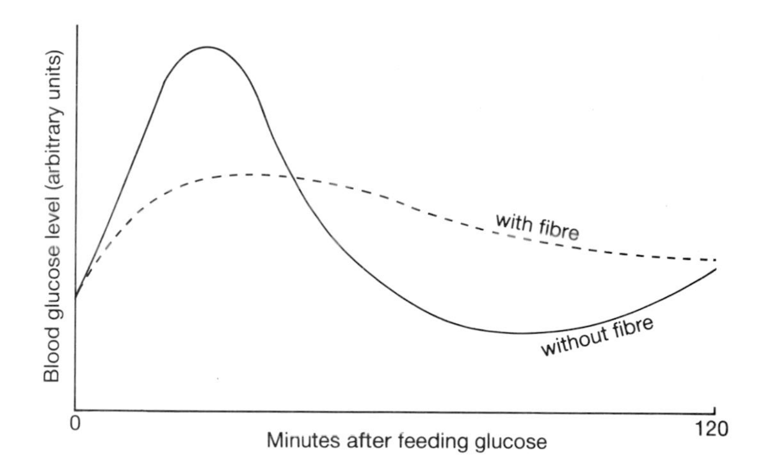

The amount of fibre in the diet can have a significant effect on the rate at which glucose is absorbed into the bloodstream after a meal. This effect can be put to good use in the management of diabetics who have difficulty controlling their blood glucose level.

of heart disease (Chapter 9), it is difficult to sort out the wheat from the chaff in these statistics.

The story of fibre merits a little philosophical reflection. No one, in particular no sector of the food industry with vested interests, can argue that fresh fruit, vegetables and whole-grain cereals are anything but good for us. And there appears to be sound clinical evidence and epidemiological arguments to suggest that these foods will actually do us some good. The interest engendered in fibre has created a demand for products such as wholemeal bread and the baking industry, or parts of it, has responded with the production of the sliced wholemeal loaf. Others have been equally innovative, to wit the vege-burgers in supermarket freezers. That represents the happy marriage between nutrition education, the food industry and twentieth-century convenience. Regrettably, this isn't always the case. The experts disagree, the food industry is divided in vested interests and the consumer suffers. At least this time, someone is happy that the greengrocer is now the local health food shop.

Summary

- The declining intake of fibre in industrialized countries has been linked epidemiologically with such diseases as diabetes, obesity, hypertension, heart disease and diverticular disease.
- Dietary fibre is derived solely from plants, particularly from the structural components (stems, seed coats) and leaves.
- Fibre is composed of cellulose, hemicellulose, pectin and lignin, the exact balance varying between foods.
- Dietary fibre is indigestible by the enzymes secreted into the small intestine and is fermented by bacteria in the large intestine.
- This fermentative process increases the microbial biomass of the large gut, leading to bulkier, softer faeces and to some flatulence.
- High-fibre diets are now recommended for diabetics because of the ability of dietary fibre to slow down the release of glucose into the blood.

4
The body beautiful

When Carlos Lopez crossed the finishing line at the end of the 1984 Olympic marathon, he had expended approximately 2900 kilocalories of energy over and above those he would have expended had he simply spent 2 hours 9 minutes watching the event on television. The energy spent in completing the 26.2 miles is obvious. That spent watching TV is less so. Nonetheless they are both based on the same metabolic process of extracting energy from food, utilizing it or storing it for subsequent use. This chapter is about energy metabolism and its defects which manifest themselves as obesity and anorexia nervosa. We begin by discussing energy in food and then the fate of this energy at the various stages of digestion and metabolism.

It's a sunny summer's afternoon and you're in your back garden enjoying the effects of a few glasses of wine. Staring into the white-hot charcoal on your barbecue you reflect, as happens with afternoon drinking, on, among other things, the amazing heat emitted from this fuel. Later in the afternoon you let a sausage fall into the charcoal which had previously charmed you. Failing to extract it (because by now the wine has made you both incompetent and nonchalant), you let it burn itself out. At your next barbecue you may again be mesmerized by the burning charcoal but you will have forgotten that the sausage was just as effective an agent of energy. All foods contain combustible material, mainly as carbohydrates, proteins, fats and fibre. Like charcoal, food can be combined with oxygen to release heat, and to give off water and carbon dioxide. The amount of heat released is measured in calories or, more correctly, as kilocalories. The 'kilo' is dropped for convenience in layman's language but I shall use the correct term. If one gram of protein is fully combusted under laboratory conditions, it will yield 5.6 kilocalories of energy. The same amount of fat and carbohydrate will yield 9.2 and 4.2 kilocalories, respectively. Fortunately, the biological combustion of nutrients is nowhere near as brutal as that of barbecue fuel or sausages therein. To begin with, not all the food eaten is digested and absorbed. As was seen in the previous chapter very little fibre, a highly combustible material, is digested and, furthermore, the efficiency of digestion and absorption of protein, fat and carbohydrate is less than 100%. This is outlined in the figure opposite, which is basically the sequence I will follow to define energy metabolism. The digested foods are absorbed mainly as glucose, amino acids and fatty acids, as outlined in the preceding chapters. While glucose and fatty acids can now be considered for combustion to yield energy (should that be what is needed), there is a slight problem with amino acids in that their metabolism gives rise to ammonia which, if permitted to accumulate, would be toxic. So, as quickly as it is formed, the ammonia is detoxified into a safe compound known as urea, which is the major component of urine (and, incidentally, explains why a bucket of soaking nappies stinks of ammonia, to which the urinary urea reverts, through bacterial action, on standing). Urea is a combustible compound and hence represents a loss of energy from the body.

There is a second possible loss of energy into urine. When the body is engaged in combusting fat, it requires some carbohydrate to assist it. If this carbohydrate (glucose) is not available, then fat combustion is incomplete and gives rise to an elevated level of ketones in the blood. These are a potential source of energy which spill over into the urine for excretion, hence boosting urinary energy losses. This is often deliberately exploited by slimmers who go on a carbohydrate-free diet, so forcing fat combustion to proceed incompletely, leading to increased urinary energy losses as

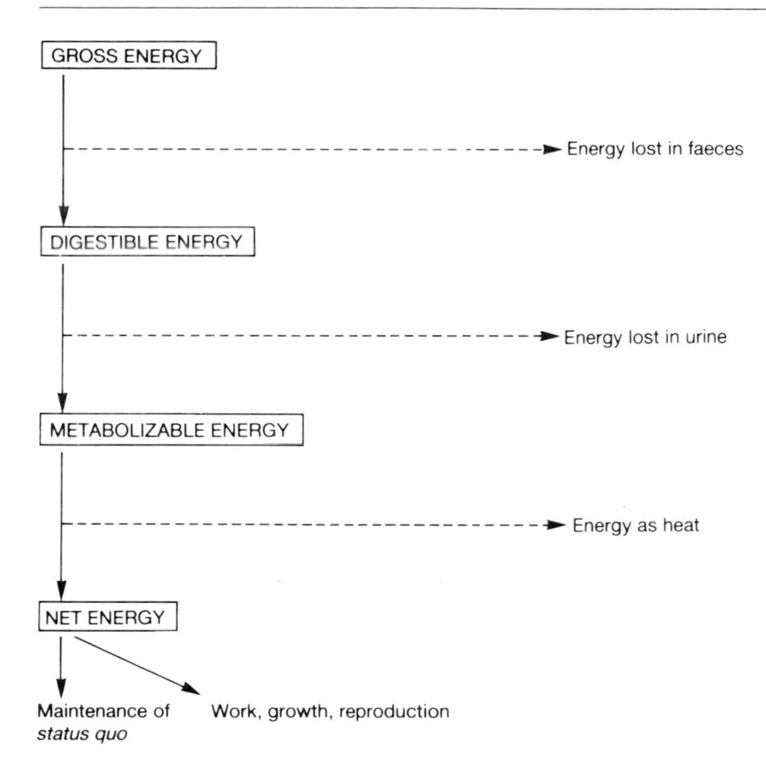

The fate of energy in the body.

ketones. High levels of blood ketones are dangerous and those pursuing a carbohydrate-free diet should be careful.

Table 4 shows the comparison between the energy available in the laboratory and what is, so far, available biologically, following deductions of what is lost in the urine and faeces from what has been eaten.

Table 4. *Energy available (kilocalories per gram) from protein, carbohydrate and fat in laboratory and metabolic combustion*

	Protein	Carbohydrate	Fat
Laboratory	5.6	4.2	9.2
Metabolically available	4	4	9

The bottom line of Table 4 represents the 'Calories' which we count, so this will allow us to calculate the calorific value of a food, say a sausage:

		3 sausages, total 100 grams	
Protein	=	13 grams, at 4 kilocalories/gram =	52
Carbohydrate	=	15 grams, at 4 kilocalories/gram =	60
Fat	=	18 grams, at 9 kilocalories/gram =	162
Water	=	54 grams, at 0 kilocalories/gram =	0
		TOTAL =	274 kilocalories or Calories

So three sausages contain 274 kilocalories and that is how nutritionists work out the Calories in a food, a meal or a nation's diet. These kilocalories are now available for the involuntary biological functions such as heart-beat, lung movement and kidney function, and for the voluntary biological functions such as walking, reproducing or swimming. Before moving on to these aspects of energy metabolism let us consider very briefly the biological extraction of energy from food into usable forms. That we inhale oxygen and exhale carbon dioxide tells us that the process is, in principle, similar to that for the combustion of fuel such as charcoal. There are, however, two features of biological combustion which distinguish it from the barbecue. The first is that the process takes place step-wise with the slow gradual release of energy to particular tissues. The second is that the process is inefficient. Only 68% of the potential energy of glucose which enters a cell is extracted as biological energy; 32% is lost as heat which helps us maintain our body temperature at 37°C.

Let us now consider our Olympic marathon runner some 4 hours before the start of the race. Breakfast consists of toast and coffee. The starch in the toast is digested and absorbed as glucose. He is resting in his armchair, psyching himself up for the big event. He has as yet no need of energy for physical activity as he sits in an apparent trance. The glucose is directed by the hormone insulin into two stores of energy, fat in adipose tissue and a substance called glycogen in muscles. I have said little of glycogen, so perhaps a short explanation is needed. Glycogen is a carbohydrate not unlike starch and acts as a local reservoir of energy for muscular activity. In

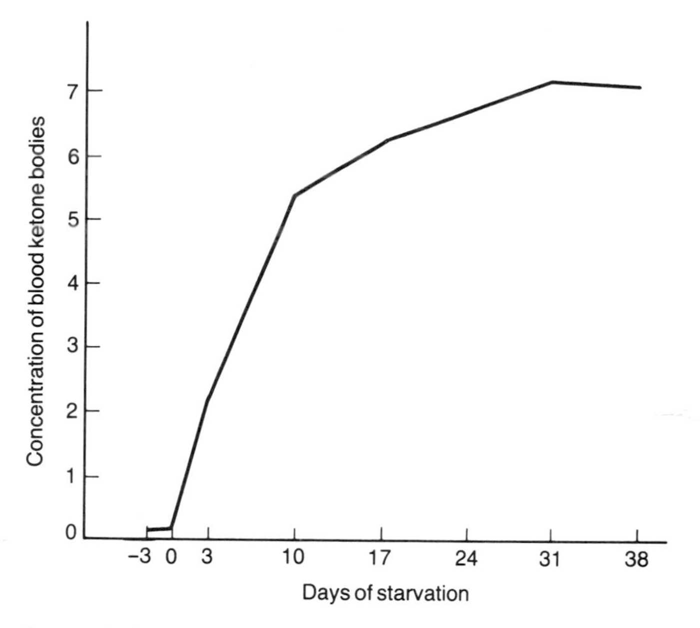

Starvation results in raised level of ketone bodies in the blood – as this graph illustrates. In the absence of carbohydrate (glucose), fat combustion is incomplete and inefficient and results in a build up of ketones in the blood – some of which is lost in the urine.

The use of this carbohydrate-free diet does however need to be carefully monitored since high levels of blood ketones are potentially dangerous.

the sedentary among us there is about 0.5–1.0 grams of glycogen per 100 grams of muscle. Athletes can, with training, increase this to about 4 grams per 100 grams of muscle, but not more. There is a very definite upper limit on this store of energy.

Fat, on the other hand, is a longer-term energy store which has a habit of expanding to whatever we dictate by diet. It has, as you may have noticed, no upper limit and clearly people vary greatly in how much fat they carry about. Alberto will begin to burn off his meagre fat stores shortly after commencing his run and by the twenty-sixth mile will be relying solely on fat energy. It is essential, therefore, to grasp the point that there are two sides to the energy metabolism coin: one for storing energy and one for disposing of that energy. It is the balance between storage and disposal which makes Laurel a lesser man, in gravimetric terms, than Hardy.

Physical activity is one route of dispersal. We shall now consider the other routes.

Involuntary energy disposal

Maintenance

One of my earliest teachers of biology used the motor car as an analogy for the energy processes of the body. There are similarities: both require fuel to function and this fuel is burned to provide energy in a slow, highly controlled process, exactly as needs dictate. Both are also inefficient in this respect. But there are outstanding differences between the car and the body. A parked car does no work, but the moment a human does no work, funeral arrangements can be made. For even in sleep our organs and tissues grind away: our hearts beat; our lungs heave to and fro; our brains remember to wake us up, provide us with dreams and a recollection of the past; our kidneys filter blood and our bladders distend with collected urine; our blood swirls to and fro! Life is the ultimate 'on-going situation', and all of these involuntary bodily functions are energy-requiring processes, 24 hours a day, 7 days a week, till death us do part. The energy spent in this way is correctly termed maintenance energy. It represents a substantial proportion of daily energy expenditure*, being largely determined by the individual's weight and body fat as shown in Table 5.

There are three points to note in Table 5. The first is that for a given body weight, fat people expend fewer kilocalories than lean people. The reason for this is that fat contributes very little to maintenance energy. It doesn't quite just sit there hanging loose and doing nothing other than depress some of us; it does turnover and metabolize, but its metabolism is, nonetheless, negligible in energy terms. In fact about 80% of maintenance goes on in the obviously active organs, the liver (27%), the brain (19%), the heart (7%), the kidneys (10%) and the muscle (20%). The skin, the digestive tract and the lungs are all also very active in the metabolic sense. Evidently, fat contributes very little.

Secondly, you may have noticed that within a given category, say thin males, the greater the body weight, the greater the energy requirement for

* An average value for maintenance energy is 1500 kilocalories per day.

Table 5. *Variation of maintenance energy (kilocalories per minute) with sex and body weight*

	Body weight (kilograms)		
	50	60	70
Thin male	0.99	1.12	1.26
Fat male	0.84	0.98	1.11
Thin female	0.89	1.03	1.16
Fat female	0.79	0.88	1.02

maintenance. Clearly, if thin people of different weights have the same proportion of fat, the extra weight represents a bigger muscle mass, bigger guts, livers, etc. Hence, a bigger energy requirement.

Thirdly, you will see that for a given body weight, say 60 kilograms, thin men expand more energy on maintenance than thin women (1612 *versus* 1483 kilocalories per day). This is because whereas a thin man's body contains about 10% fat, a thin woman's body contains about 20% fat! The comparable values for fat people might be 20% and 30%, respectively. How come? Well, God in His Wisdom has endowed women with more fat than men and I don't intend to aggravate the feminists among you by suggesting that the extra fat is to ensure a reserve of energy for the most demanding of physiological functions, pregnancy! I'm just grateful that the Lord in His Wisdom chose to distribute this fat as he did: about the breasts, thighs and pelvic region making the female form more aesthetically pleasing than the male. And before any of you feminists have a seizure, let me tell you this. This additional fat is not only of benefit as an energy reserve in childbearing, however, it is equally useful wherever energy supplies are at risk: among explorers, spies, soldiers, submariners, astronauts and other such male-dominated professions. Furthermore, this extra fat permits better adaptation by females to colder climes and doesn't hinder their equally superior physiological adaptation to heat.

Returning to the energy for maintenance, you can see from Table 5 that a value of just over one kilocalorie per minute is a reasonable average, which in a 24-hour period amounts to about 1500 kilocalories.

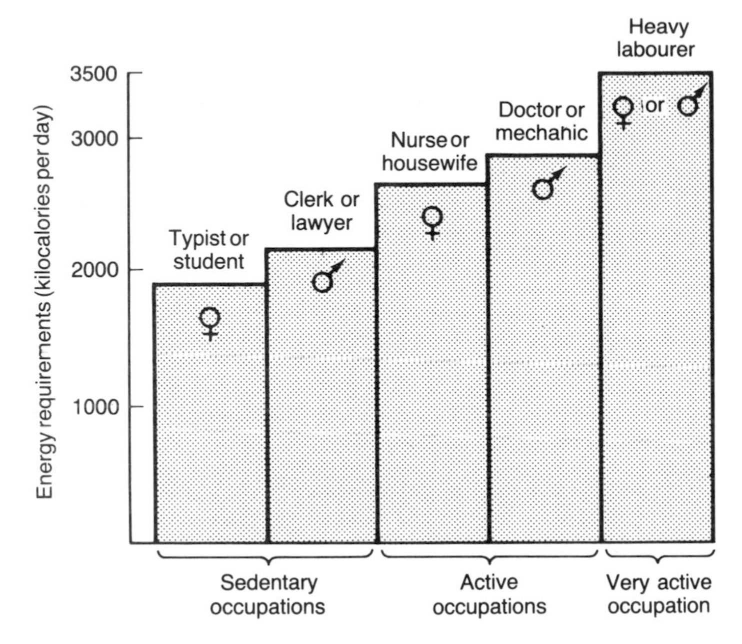

Although a person with a very active lifestyle does need more energy than someone with a very sedentary existence (as shown in this bar chart), it is surprising that a substantial proportion of our energy requirements is used simply to keep the body ticking over.

The maintenance of body temperature

To this day abandoned newborn infants, especially at Christmas time are worthy of headlines. How such infants survive in the cold never fails to amaze us. They do so by a process known as non-shivering thermogenesis. The term thermogenesis is very widely used to discuss the utilization of food energy by the body specifically for heat production (or in some cases, the production of heat which is only secondary to some other physiological process such as physical activity). Non-shivering thermogenesis, as the name implies, involves the physiological adaptation to cold without resort to shivering. In infants this is the only protection against cold. In adults, however, shivering thermogenesis is the first obvious reaction to cold, to be replaced gradually by non-shivering thermogenesis as the individual adapts to the cold environment.

Diet-induced thermogenesis

Calorimeters are devices for measuring heat production. Such a device is used to measure the kilocalories of heat released on combustion of samples of food or fuel in the laboratory. There are in existence many calorimeters the size of modest living rooms. These staggering contraptions can be used to determine the heat produced by the human body at work, rest or play.* Let us now consider a volunteer sitting in just such a calorimeter, contemplating little but his or her exposed navel. The digital display indicates a heat loss by the subject of about 1.1 kilocalories per minute, which is what you would expect because in this rested state the heat loss is almost totally dominated by the maintenance component. In other words the individual is maintaining life and we are measuring the heat produced by the millions of inefficient reactions going on in the process. Now the subject takes a glass containing a liquid meal, such as is commonly available for convalescents, and drinks its contents, equivalent to an average meal of say 500 kilocalories. We wait. Gradually the digital read-out rises, indicating a detectable increase in heat loss. It continues to rise and peaks within about 40 minutes, subsiding as slowly back to basal levels. This heat is known as diet-induced thermogenesis. It probably arises from the multitude of metabolic activities following the meal: digestion, absorption, transport and metabolism. None of these heat-producing events would have arisen had the subject remained fasted and they are consequently specific to the uptake and metabolism of the meal. Nonetheless, had the subject not eaten the meal, he would have continued to produce some heat because of the metabolism needed for maintenance, which continues whether he eats or not. Metabolism, whether it be for maintenance or for assimilating the components of a meal, is energetically inefficient and so produces heat.

The amount of extra heat, above maintenance level, that is lost to the environment following a meal (diet-induced thermogenesis) varies according to the type of meal and, possibly, according to the individual. This is a point to which I will return shortly but suffice it to say for the present that if

* This is known as direct calorimetry, although most estimates have in fact been made using indirect calorimetry.

(*a*)

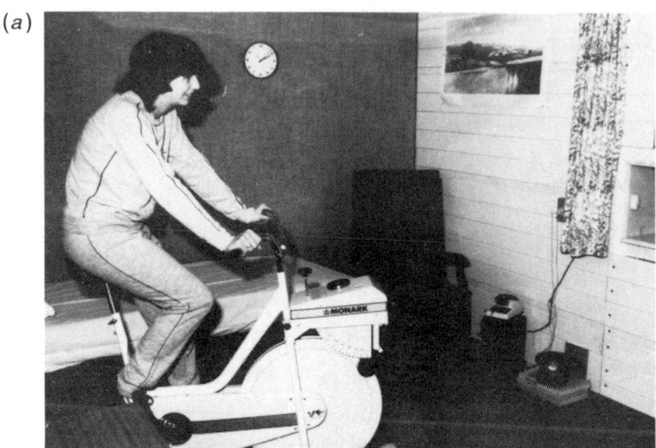

(*b*)

The inside of this room-sized calorimeter might look much like any other room (*a*). However, on the outside (*b*), a wide range of sophisticated equipment is needed to measure and monitor the changes in heat loss that occur within the calorimeter. (Photographs courtesy of the Dunn Nutrition Unit, Cambridge.)

two people eat identical meals and one manages to produce more heat energy as diet-induced thermogenesis, then it follows that that person has less left to deposit as fat. Whether this is applicable to human nutrition is highly contentious.

The voluntary expenditure of energy

We have little control over the loss of heat from the body via shivering, non-shivering thermogenesis, the thermogenesis associated with mainte-

Table 6. *Cumulative weight gains (grams) in different tissues during pregnancy*

	Duration of pregnancy (weeks)			
	10	20	30	40
Foetus + placenta	55	720	2530	4750
Uterus + breasts	170	765	1170	1300
Blood	100	600	1300	1250
Water	neg.	neg.	neg.	1200
Fat	325	1915	3500	4000
TOTAL	650	4000	8500	12 500

neg. = negligible

nance of metabolism, or diet-induced thermogenesis. There are, however, some forms of thermogenesis and energy output that we can regulate ourselves. We can generate heat by exercise and, within reason, most females can choose to become pregnant if they so wish. These are the next two topics for consideration.

Pregnancy and lactation

The gain in weight during an average pregnancy is categorized according to tissue and time in Table 6.

In the last 10 weeks, the foetus and placenta almost double in weight and finally account for the biggest component of the gain in weight. In the penultimate 10 weeks, the gain in fatty tissue almost doubles. Together these tissues comprise 70% of the 12.5-kilogram gain. The total amount of energy needed to produce an infant is calculated to be about 80 000 kilocalories: 7500 for the baby itself; 7500 for the placenta and reproductive organs; 40 000 for the extra fat stores; and 25 000 for the increased maintenance of all the newly formed tissues. After delivery there is a slow return to normality, particularly so if the mother breast feeds, for this is the surest way of dissipating that 25 000 kilocalories of fat deposited during pregnancy. The daily cost of lactation is about 750 kilocalories which, coupled with a normal intake of energy over a nursing period of 4–5 months, will soon lead to a normal body composition.

Stamina Rating	
Badminton	★★
Canoeing	★★★
Climbing stairs	★★★
Cricket	★
Cycling (hard)	★★★★
Dancing (ballroom)	★
Dancing (disco)	★★★
Digging (garden)	★★★
Football	★★★
Golf	★
Gymnastics	★★
Hill walking	★★★
Housework (moderate)	★
Jogging	★★★★
Judo	★★
Mowing lawn by hand	★★
Rowing	★★★★
Sailing	★
Squash	★★★
Swimming (hard)	★★★★
Tennis	★★
Walking (briskly)	★★
Weightlifting	★
Yoga	★

★ not much effect
★★ beneficial effect
★★★ very good effect
★★★★ excellent effect

Regular exercise can play an important role in the maintenance of good health, as well as being an aid in controlling weight. This diagram rates the beneficial effects to be gained from various types of activity. Jogging, swimming, cycling and rowing all provide the highest stamina ratings (data reproduced with permission of the Health Education Council).

This, incidentally, is the first of several advantages of breast feeding over bottle feeding. The others will be dealt with elsewhere.

Exercise

Ours is a lazy civilization of escalators, elevators, automobiles, desks, automatic washing machines, and so forth. The pressure on us to get off our butts is considerable. Picture the individual who decides to walk a mile home or to the railway station, in a sop to this pressure. The extra energy expended over and above what would have been spent driving or sitting on a bus is 50 kilocalories, equivalent to about one apple. If the same individual were to take up jogging, say three times per week over a modest 6 miles, the extra energy he or she would expend would be about 325 kilocalories for each of the 7 days of the week. In general, the expenditure of energy above maintenance for jogging is one kilocalorie per kilogram body weight per kilometre run; for me, this is 3500 kilocalories per marathon. You can work out your own. Swimming can be generally regarded as four times less efficient in terms of energy expenditure than running, which means an energy cost of about 4 kilocalories per kilogram body weight per kilometre. If, like me, you run four times faster than you swim, the easy way out is simply to talk of 12 kilocalories per kilogram body weight per hour of intensive exercise. From these values it is obvious that we can exert a considerable effect on the expenditure side of the energy coin through exercise, although the true benefits of exercise have little to do with body weight regulation.

Energy balance and obesity

The average non-pregnant, non-lactating person consumes about 2400 kilocalories per day. Of this, about 1500 kilocalories is required for maintenance, leaving about 900 kilocalories to be disposed of through shivering, diet-induced thermogenesis or physical activity. Most of us encounter no particular problem in dissipating this remaining 900 kilocalories, since most of us maintain a stable body weight for pretty long periods. Some, however, put on weight and become fat or obese. The cause of this obesity is elegantly hidden within such statements as: 'obesity develops when the expenditure of energy is less than the intake of energy'. One bone of contention is whether obesity develops because people eat above-average levels of kilocalories or because they eat an average amount but can't for some reason dissipate it all. The following two quotations embrace the controversy surrounding this issue.
Favouring an exessive intake:

> There is also the perennial problem of how some of us appear to eat like horses and remain thin, whilst others, if they do not continually exercise conscious restraint on their food intake, readily put on excess weight. It is postulated that the

Total exercise time = 4 mins (240 secs)

1st pulse reading = ... beats in 30 secs
2nd pulse reading = ... beats in 30 secs
3rd pulse reading = ... beats in 30 secs

Total pulse reading = ... beats
(add up your 3 readings)

Multiply your Total pulse reading by 2 to get your Grand pulse reading

Grand pulse reading = ...

$$\text{Fitness score} = \frac{\text{Total exercise time}}{\text{Grand pulse reading}} \times 100$$

= ....

Fitness scores:

over 90 — very fit indeed
81 – 90 — very good
71 – 80 — good
61 – 70 — quite good
51 – 60 — poor
under 51 — very poor

One simple way of quantifying fitness is to perform the following exercise. Spend 4 minutes alternately stepping onto (with both feet), then down from, a box about 40–50 cm high. Do this at a steady pace, say one step up every two seconds. Then sit down and rest for 1 minute exactly; then measure your pulse for 30 seconds. (1st pulse reading.) Rest for a further 30 seconds; then measure your pulse for the next 30 seconds. (2nd pulse reading.) Finally, rest for a further 30 seconds, then take your 3rd pulse reading for the next 30 seconds. Use the table to calculate your fitness score.

thin people keep thin because they operate thermodynamically inefficient metabolic pathways, or futile cycles, and that these are concentrated in brown fat. I have no enthusiasm for these views.

Favouring a metabolic defect:

> There are obese people who eat large quantities of food while others apparently have normal food intakes and some actually eat less than their lean counterparts. It has so far proved very difficult either to quantify or explain these differences but it is clearly no longer justified to assume that obesity is a self inflicted disease which should not receive medical attention.

You would think that the solution to this problem would be simple. Take a group of normal people, overfeed them and see what happens. If excess food intake is the be all and end all of obesity, then these non-obese volunteers should put on weight. Between 1902 and 1977, some thirteen such studies were carried out. Eleven of these would lead one to conclude that when we overeat, for quite long periods, we still manage to maintain a reasonable body weight, that is, the variable weight gain which occurs is nowhere near what would be predicted from the extra kilocalories consumed. So where is the extra energy going? Is there a special mechanism for dissipating unwanted calories and is a defect in this mechanism the basis of much of the obesity seen today?

One mechanism under extensive investigation these days is the role of brown fat in increasing diet-induced thermogenesis, so that following a meal there is an above-average amount of heat lost by the body, to compensate for the excess kilocalories consumed. Brown fat is most abundant in small mammals, especially those which hibernate, and in young infants. In fact it is brown fat which is responsible for the non-shivering generation of heat in those abandoned waifs, and it works in the following way. Normally, one molecule of glucose or fat is combusted within the cell in a well-ordered manner; granted it is inefficient, in that only 68% of the potential energy is trapped. In brown fat, the combustion of glucose and fatty acids is woefully inefficient, deliberately so, leading to a staggering reduction in the biological entrapment of energy. It follows that if an individual can switch on his or her brown fat as needed, then, within reason, stable body weight can be maintained. The difficulty lies in identifying brown fat in human adults for, frankly, it is difficult to find and, having found it, show it to be active. Hence the scepticism about brown fat

Table 7.

	Overeaters	Undereaters
Age (years)	25	31
Weight (kilograms)	71	81
Height (metres)	1.83	1.76
Percentage body fat	12.6	22.5
Usual energy intake (kilocalories per day)	3586	1650
Percentage rise in resting metabolic rate on eating a meal	22–29	8–20

and the role of diet-induced heat production in the maintenance of stable body weight. Even so, it is possible to see how it might work from the following experiment carried out at the University of Southampton's Department of Nutrition. Two groups of volunteers were recruited, one group was classified as 'undereaters', the other as 'overeaters'. You can see from Table 7 that the overeaters were individuals who were lean and ate a lot. Both groups were given the same test meal and energy output was measured in a resting position following the meal. The increased energy output of the leaner group was 250% higher than that of the fatter group. The lean volunteers could eat to their heart's content and not gain weight, simply by burning it off. Is brown adipose tissue the crucial factor?

Not everybody would agree with these results but I must draw the line here and reveal my prejudice in favour of diet-induced thermogenesis as a means of regulating normal body weight and, by implication, in favour of a defect in this mechanism in a considerable proportion of the grossly overweight. Finally to find out which side of the fine divide you lie, you can consult the graph showing optimum weights for people of different height. Draw a line vertically upwards from your height and horizontally from your weight. Where they intersect is the point you're looking for. Hopefully it's between the lines, preferably nearer the lower one!

Anorexia nervosa

Not all the problems of weight regulation involve obesity. Today there exists the phenomenon of anorexia nervosa, a disease which appears to be most prevalent in adolescent females, especially among the middle classes. It appears to be derived from a desire to slim through dieting which, for some psychological reason, develops into a distorted obsession with the

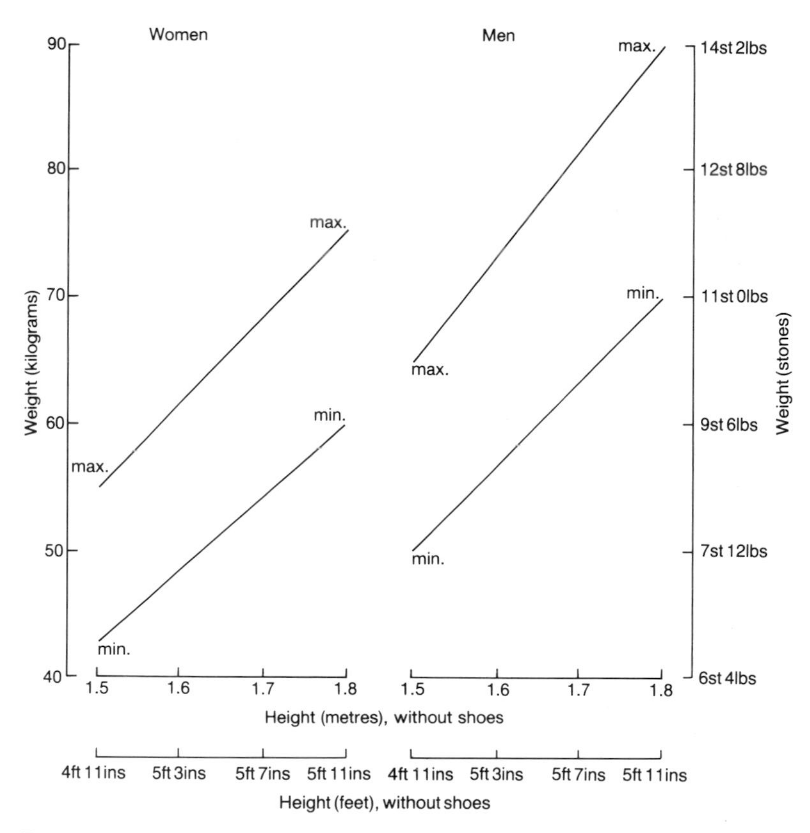

Range of preferred or desirable weights for men and women wearing indoor clothing. (Based on a report of the Royal College of Physicians on obesity.)

shape of the body. It is primarily a psychological disorder, to be psychologically treated, and is not at all well understood.

Summary

- Energy is measured by most lay people in Calories, although strictly speaking it should be kilocalories. This energy is derived in the body from protein, fat and carbohydrate.
- After losses of calories in faeces, urine and body heat are allowed for, the energy obtained from one gram of carbohydrate, fat and protein is 4, 9 and 4 kcal, respectively.

- There are several routes by which food energy is expended; these are either voluntary or involuntary.
- The maintenance and functioning of essential organs is the first prerequisite for energy, i.e. for the heart, lungs, liver, gut, brain, kidneys, etc. Other involuntary outlets include growth, the heat generated on eating, and the maintenance of the body's temperature.
- Voluntary energy outlets include pregnancy and lactation and, especially, exercise which is the one route of energy outlet which we can increase most easily.
- There is, as yet, considerable controversy as to whether obesity is simply an excess of energy intake over output, or whether it involves a partial defect in energy expenditure such as in the generation of heat after eating in individuals consuming 'normal' levels of calories.

5
How the other half dies

Within one decade no child will go to bed
hungry, no family will fear for its next day's
bread, and no human being's future and
capacities will be stunted by malnutrition.
Every man, woman and child has the inalienable
right to be free from hunger and malnutrition . . .

World Food Conference, Rome, 1974

Whilst we in the developed world worry ourselves about thermonuclear war and the diseases of affluence, cancer and heart disease, an embarrassing proportion of our brethren in the less-developed countries grapple daily with diseases arising directly from an erratic food supply. The problem pricks our conscience sufficiently into establishing and funding charitable agencies and, during quiescence of famine, we somehow feel that in so doing we have discharged our duty to this dreadful global problem. Then like a bad wart, the ugliness erupts, precipitated by drought, floods, earthquake, pests or war, to a level sufficient to merit the dispatch of the television crews to glaze our evening news with haunting pictures of emaciated infants, bereaved and wailing mothers and hoards of witless refugees. This blight will persist and continue to reappear periodically until a real solution is found, one along the lines of the Brandt Commission's report, requiring permanent shifts in the distribution of the earth's resources between north and south. However, whilst an emotional approach might put us in the right frame of mind to appreciate the issues, a logical approach is the only way to understand the complex issues involved and see some hope for progress.

Table 8. *Daily energy supply per head as a percentage of daily per capita requirement*

	1961–3	1964–6	1969–71	1972–4
Developed economies	125	126	131	134
Seriously underdeveloped economies	91	90	92	90
World average	101	103	106	107

There was a time, not all that long ago, when the prevailing wisdom was that the growth of the world's population had outstripped its food supply, requiring a technological solution – novel proteins from waste petroleum products or from sewage by-products. The facts are, however, that there is enough food available were it distributed more equally. Table 8 shows the world energy supply expressed as a percentage of the world energy needs. Clearly, these figures are only relevant in the short term and will need to be reviewed with rising global population. There is, however, no justification, moral or economic, for energy-consuming solutions such as some of the world's major chemical corporations might have us believe.

A second important preliminary observation is that even within a given country there is great regional and socioeconomic variation in malnutrition, otherwise known as the 'Beverley Hills is not Harlem' law. The urban poor and rural peasants are the two groups most at risk. The indigenous middle and ruling classes are generally immune for a variety of reasons, not least because they call the shots, economically and militarily speaking. Recife and São Paulo, two Brazilian cities, provide an example of regional variation. The death rate in preschool infants (1–4 years) per 1000 inhabitants in Recife (9.0) is almost three times higher than that in São Paulo (2.8) which, in turn, is many times higher than that in the developed West. Socioeconomic factors are again evident from statistics gathered in Kampala, Uganda: among the severely malnourished, a mere 8% were born to homes with a salaried income. Against this political and economic background, let us now consider the nuts and bolts of malnutrition.

There are three categories of malnutrition: marasmus, wasting and kwashiorkor. These diseases are not so pathologically distinct as, say, polio, measles and diphtheria. There is a great deal of overlap, so that a child may experience aspects of all three forms of malnutrition.

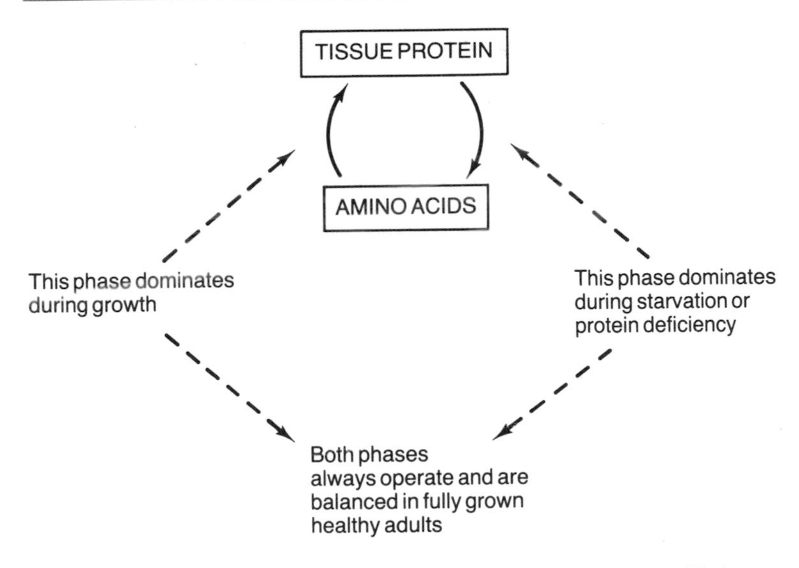

For those on inadequate protein diets, the right-hand side of this equilibrium predominates with tragic results.

The utterly emaciated child with long bones protruding through an outsize skin and sunken staring eyes, the most haunting feature of advanced famine, is suffering from classical marasmus. This is a medical term for down and out starvation. In the preceding chapter, I outlined the concept of resting metabolism, the obligatory expenditure of energy by the essential processes of life. If caloric intake falls below this basal level, we lose weight. Every would-be-slimmer knows that. In infants the same applies, except that in infants there is the further demand for the energy for growth. Therefore, an inadequate intake of energy immediately leads to an arrest of growth. As the deprivation continues, body weight, slight as it might be, begins to dwindle. Eventually death ensues. Well-built adult prisoners in Belfast's H-Block survived for about 60 days of starvation. Infants don't fare as well. It must be a dreadful death.

The shocking thing is that this emaciation is seen as a physiological 'adaptation' to starvation, for it is a fact that some tissues are more vital to life than others. Muscles are useful. Big muscles may suit pentathletes but starving children have more need for digestive enzymes, antibodies and a functioning liver than an attractive physique. So as proteins turn over in muscle, their constituent amino acids are shunted away and used for the synthesis of more vital proteins and tissues. The marasmic child has lived off its own muscle to survive caloric and protein deprivation. The child with kwashiorkor, as we shall see, has failed in this adaptation with perilous consequences.

Marasmus may be the clarion call for news coverage. A vast number of children live dangerously close to marasmus, however, although not close enough to merit this distinguished diagnosis. They are classified as suffering from muscle wasting. The distinction is quantitative. Marasmus is 'defined' as a decline to 60% of the expected weight for a given age, while wasting is described as a body height which is less than 90% of the ideal height for a given age.

In fact, wasting is the most prevalent form of malnutrition encountered in Third World countries. Clearly it would take little by way of social, domestic economic, political or health upset to push 'wasted' infants further down the ladder of malnutrition towards marasmus. Hence the rapidity with which marasmus can occur. Nonetheless, children who suffer from wasting, and indeed those who have suffered marasmus and survived, will exhibit the phenomenon of catch-up growth upon suitable nutritional rehabilitation. This is characterized by a period of accelerated growth which occurs upon refeeding following a period of malnutrition. A normal infant will eat 120 kilocalories a day and gain 3.3 grams. An infant in catch-up growth will eat 170 kilocalories per day and gain 10 grams. Cattle foraging the bare pastures of winter enter spring lighter than they left autumn but quickly compensate for this deprivation and rapidly catch up with their target weights. So too do children who are rehabilitated after malnutrition. But their catch-up growth is often incomplete, depending on the extent and severity of their undernutrition, leaving them permanently smaller than would be expected for their age. In every other physical respect they are normal. This is known as stunting. The callous question now arises as to whether stunting *per se* is a condition requiring treatment or whether, as it has been aptly put, it represents no more than a scar of having suffered and survived. What, you may well ask, is problematic about being smaller than the West's concept of expectation? Could it even be regarded as a further manifestation of adaptation – a smaller mass of cellular tissue to feed? There may be good arguments for so regarding stunting but there remains the possibility that smallness, or stunting, might be associated with impaired intellectual performance, leading to social and behavioural handicaps. That could close the vicious circle of malnutrition.

The development of the human brain begins in gestation and continues well into the second year of life. It is a dual development, involving an increase in both the size and the complexity of the brain. No one can truly answer whether prolonged nutritional deprivation during that critical period of growth will have a lasting deleterious effect on brain function. Animal studies suggest this possibility and, for as long as they do (in spite of different patterns in brain development between animal species in the pre- and postnatal periods) there remains the possibility of impaired intellectual performance in stunted children.

One should not attempt, however, to attribute everything to diet, for diet is only one aspect of a deprived culture. Children with malnutrition tend often to be the children of poor parents or to hail from 'broken' homes. They tend to be sick, to miss school or never attend. So impaired intellectual performance is more likely to be a sociological phenomenon which parallels malnutrition. One author has written thus:

> The light of curiosity absent from children's eyes. Twelve year olds with the physical stature of eight year olds. Youngsters who lack the energy to brush aside the flies collecting about the sores on their faces. Agonisingly slow reflexes of adults crossing traffic. Thirty year old mothers who look like sixty. All are common images in developing countries; all reflect inadequate nutrition; all have societal consequences.

As the late Mrs Indira Gandhi commented on the poorer regions of India: 'It is quite common to be told that the people are lazy and indolent'.

The third category of malnutrition is the most complex and the most evil. It is known as kwashiorkor. The infant, unlike the bag-of-bones syndrome of marasmus, is chubby: chubby cheeks and eyes; chubby limbs and belly. It is a 'sugar-baby', except that the chubbiness is not due to fat but to the accumulation of water under the skin, otherwise known as oedema. Underneath that puffy skin lie other anatomical abnormalities – an enlarged fatty liver and a porous, badly functioning gut. Further chemical investigations reveal changes in blood chemistry, central to which is a low level of the blood protein albumin. Such analysis also reveals

CAMROSE LUTHERAN COLLEGE
LIBRARY

anaemia and low levels of circulating antibodies which protect against infection. This derangement arises because of a hormonal imbalance, which is related to inherent genetic factors, diet and infection. In the child with kwashiorkor, insulin levels are normal or raised. Consider a child who is eating a diet rich in starch but low in protein. The starch intake stimulates secretion of insulin. Insulin, as we saw above, is the hormone which directs glucose into muscle; it also directs the limited supply of dietary amino acids into muscle. So the drift is towards maintaining muscle at the expense of other organs such as digestive tract or the lymphatic tissue or the liver. If protein synthesis in the liver is reduced because of a low dietary intake and an elevated 'muscle favouring' insulin level, then key liver export proteins fall. There are two of importance: albumin, which regulates among other things the water content of blood; and lipoproteins, which carry fat, synthesized in the liver, to other organs. As a result of these changes, fat accumulates in the liver and water accumulates in tissues, producing oedema. Such a child will lose its appetite for its low-protein – high-starch diet, so that by the time it arrives at a clinic (if at all) many of these hormonal changes have shifted. The phenomenon of marasmic kwashiorkor may be seen.

It is time now to consider in greater detail the nutritional reasons underlying protein energy malnutrition. An uncomplicated shortage of food below that required for normal growth will lead to wasting. Further deprivation will lead to marasmus. An adequate or near adequate intake of a high-starch/low-protein diet is one means of altering the hormonal balance leading to kwashiorkor. But there are other factors, infection and social deprivation. Of all factors, however, weaning has the greatest impact physiologically and psychologically. In general, lactation is adequate in times of moderate nutritional stress although it will completely cease in the extreme. The composition of milk will remain relatively constant. It is an energy-rich food with one of nature's highest quality proteins. It is generally adequate in vitamins. It is rich in calcium and phosphorus for bone growth. Many of the trace elements, especially iron, are present in forms which make them almost completely available for absorption. Furthermore, it provides a reasonably important supply of antibodies to act locally in the child's gut to help stave off gastrointestinal disorders. It is nature's own infant food. Consider, therefore, the digestive shock which a

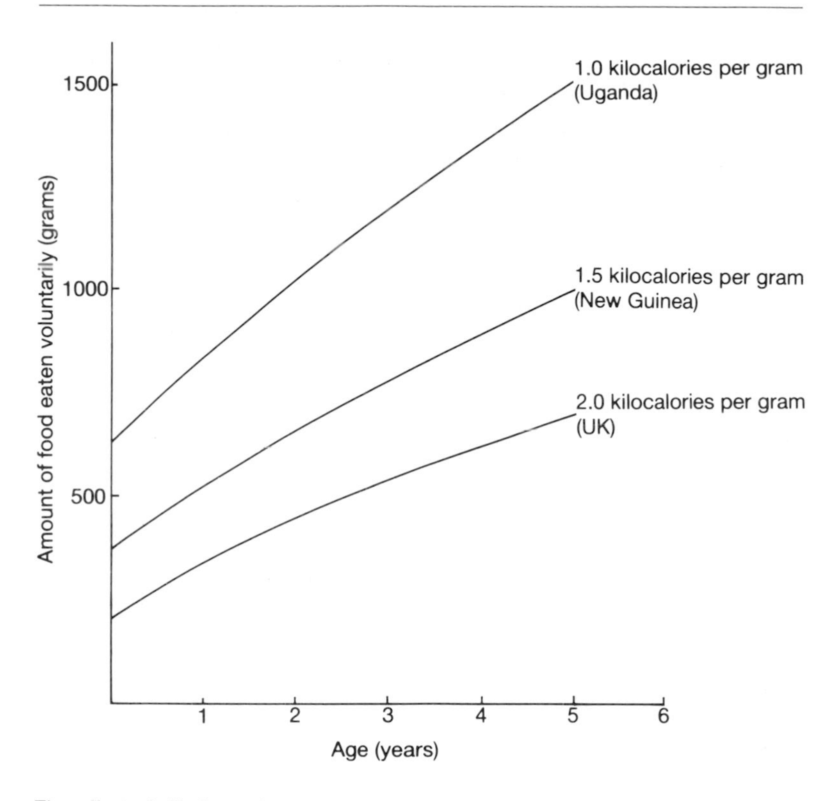

The effect of diluting calories with bulky fibre on the intake of food by children.

child encounters on reasonably abrupt weaning, not to tinned, puréed meat and vegetables or sanitized 'humanized' milk substitutes, but to low-protein, high-fibre foods of cassava or millet with perhaps an over-dilute, not-too-clean, relatively expensive artificial milk. The consequences are a sharp drop in energy intake. Even if such foods are in adequate supply, it is physically impossible to put the required starch and protein into the child by virtue of their dilution with water and fibre. (See illustration above which compares the weight per day of different weaning foods consumed voluntarily by infants.) A Ugandan child, with cassava as the staple food, must eat twice as much as a UK child to get the same intake of energy. Now consider the fortunes of such a child when food is scarce or supply erratic. Growth is immediately checked. The likelihood of infection leading to

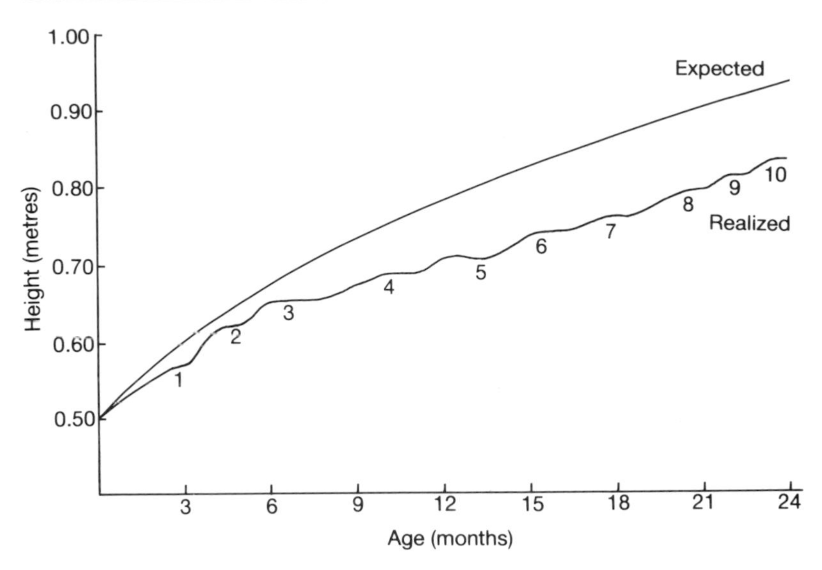

Effect of ten successive infections on growth of a young child.

diarrhoea or vomiting or both increases. Anorexia or loss of appetite results and the road to marasmus looms ahead. This is illustrated graphically above, where the growth of a child is plotted against the potential of that child for growth. Each hiccup in growth, each loss of weight is associated with one of ten bouts of illness, spanning a 2-year period, ranging from a 2-day bout of diarrhoea to a 4-week bout of bronchitis. The child may weather the storm, indeed many do, to encounter a period of wasting, managing to emerge a young preschool infant, physically unscathed except for stunting.

Clearly, the importance of infection in malnutrition cannot be underestimated. Infection causes the adrenal glands to secrete cortisol which rapidly converts muscle protein back into blood amino acids for subsequent use elsewhere – the gut enzymes, antibodies, haemoglobin, blood albumin, blood lipoproteins, etc. Children who suffer repeatedly from infection are more likely to go down the road of marasmus. It could be that those who succumb to the deadly kwashiorkor have a genetic inability to produce enough cortisol. It could be. This aspect of hormones in malnutrition is a vital area of research. If this biochemical and physiological investigation of such a horrible disease, a needless disease, seems ghoulish and repugnant, let me assure you that it has paid dividends in the prevention and early diagnosis and the subsequent treatment of what is generally described as protein energy malnutrition.

There are several approaches to understanding the onset of kwashiorkor. Diet is one. Infection is another. But why do some children develop kwashiorkor when others in the same environment, eating the same diet and succumbing to the same infections, do not? One author has described a sociological explanation from her studies in Kampala. Kwashiorkor is likely to occur between 12 and 35 months of age. Two-thirds of the children with kwashiorkor are from parents who have migrated from their cattle-grazing homelands to the city. Although the great majority of East African babies are breast fed, none of the 50 children with kwashiorkor under study was being breast fed at the time of study. Only 40% of the kwashiorkor children were from homes with the parents living together: 40% of parents were separated; the rest were unmarried, uncertain or dead. Thirty-four per cent of children with kwashiorkor were sent away from home at weaning and a further 22% after weaning. That incidence is nearly four times the normal number. And so it goes on. In this inner-city investigation of kwashiorkor, the social deprivation was undeniable.

No one pretends that resolving the problem of malnutrition will be easy. It won't be a technological solution. There is no miracle pesticide or strain of rice or fertilizer. The solution will be political. We in the developed West must face up to the fact that we monopolize and squander the earth's resources. One day we will be forced to listen to the Heaths and the Brandts. One day, Third World countries will become nuclear forces. Then we'll listen!

Summary

- At present, there are sufficient calories and protein produced globally to feed all mankind. Alas, this food is unfairly distributed around the globe and within many countries.
- There are three categories of protein energy malnutrition: marasmus, wasting and kwashiorkor.
- Marasmus is out and out starvation, such as is seen in acute famine. It is

a deficiency of weight for a given height.

- Wasting or stunting is a deficiency in height for a given age. When a period of undernutrition is followed by nutritional rehabilitation, accelerated or catch-up growth occurs. However, if the undernutrition is severe or prolonged, then catch-up growth may be inadequate. The child does not attain its rightful stature, i.e. it is stunted. Wasting may also lead to impaired intellectual development.
- Kwashiorkor is a serious disorder of malnutrition which, for nutritional reasons, produces a hormonal profile leading to an inhibition of protein breakdown. Thus, vital functions of the liver, gut and immune system fail, while protein lies trapped in the relative luxury of muscles.
- Infection is a major cause of malnutrition leading to a stop–start growth pattern.
- Early weaning from the breast, to dilute and dirty milk formulas, denies the child its rightful supply of maternal antibodies and increases the frequency of infection.
- Weaning foods in which calories are diluted by bulky fibre also lead to growth retardation and a tendency to infection.

6
Ashes to ashes

The skeleton: calcium and phosphorus

The last component of the human body to yield itself back to mother earth, whence it came, is the skeleton. Ironically, it is the most conservative tissue, with the slowest turnover. The fastest metabolizing tissues decay much more quickly.

The skeleton comprises about 3% of adult body weight. It is made primarily of calcium and phosphorus, in the ratio of 2:1 in favour of calcium. Although the skeleton turns over less rapidly than soft-tissue organs, it still undergoes constant synthesis and breakdown. In the young and growing, this turnover is rapid and the synthesis of bone dominates. The adult shows a slowing down in turnover but, nevertheless, at any time about 3% of bone is being renewed. In the elderly, the relentless turnover continues but, as the ageing process increases, the rate of breakdown dominates so much that bones become more brittle and more prone to fracture.

Besides its obvious vital function as the scaffold upon which the human body is constructed, the skeleton plays a central role in the life of the building it upholds. The level of calcium in blood is rigidly controlled by two hormones which work to maintain a concentration of 10 milligrams per

100 millilitres of blood plasma. If the level falls below this, parathyroid hormone is released and, together with vitamin D, acts to increase the absorption of calcium from the gut and to remove calcium from bone. Above 10 milligrams of calcium per 100 millilitres of blood plasma, another hormone, calcitonin, comes into play and inhibits skeletal breakdown while allowing bone formation to continue to remove calcium from the blood, thus restoring blood calcium to normal. Half of the calcium in blood is bound to proteins, the other half is free and has a positive electric charge. It is the free fraction of calcium which has to be rigidly controlled, for these positively charged particles are involved intimately in the normal rhythm of the heart, in nerve impulses, and in muscular contractions. A decrease can cause tetany and convulsions; an increase can cause coma with heart and respiratory failure. Calcium metabolism is firmly linked with vitamin D as we saw with the function of the parathyroid hormone. This vitamin also aids calcium absorption.

About 90% of all body phosphorus is found in the skeleton and phosphorus is so widespread in the diet that a deficiency is very unlikely. Phosphorus is involved in many metabolic functions, most notably in storing energy in muscle for contraction and in trapping energy when foods are being combusted biologicaly.

The disease of osteoporosis is defined as bone loss sufficient to bring about one or more fractures with minimum trauma. This bone loss is an age-related phenomenon, although clearly the degree of bone loss is not uniform in the adult population. A group of men and women in their mid-fifties to mid-sixties was monitored for bone loss over an 11-year period. Some 38% of the men and 22% of the women studied showed no bone loss at all. In women, it is the rapid loss of calcium at the menopause which most contributes to their elevated risk of osteoporosis. A variety of factors are involved in the development of osteoporosis, one of which might be habitual calcium intake in the diet. One study in Yugoslavia compared two regions, one with a high intake of calcium, the other with a low intake (1100 *versus* 500 milligrams per day). By middle age, individuals in the high-intake group showed a greater bone density than those with the low intake. In subsequent years, the age-related phenomenon of bone loss occurred in both groups and at the same rate of loss. However, those in the high-intake group started from a greater overall calcium reserve and so fared better. Increasing calcium intake when bone loss has begun will not reverse or reduce the speed of this process. The moral is: eat a balanced diet right throughout life.

Magnesium

Magnesium plays a vital function in a variety of cellular reactions, including protein synthesis. It is also involved with calcium in the activities of nerves and muscles. An average adult body contains about 30 grams of magnesium, half in bone, a quarter in muscles and the rest in cells of other organs and tissues. Because magnesium is so widespread in foods, it is unlikely that a deficiency could occur in normal healthy individuals. Alcoholism and malnutrition are associated on occasions with magnesium deficiency, but since they are also associated with a wide range of other deficiencies, we must see them as exceptional instances. Magnesium is therefore not associated with a deficiency disease and warrants little further discussion.

Iron

One of the main functions of iron is as a component of two molecules, both of which are designed to trap and transport oxygen: haemoglobin in the blood, and myoglobin in the muscle. About 2500 milligrams of iron are found in haemoglobin and 150 milligrams in myoglobin in the text-book man of 70 kilograms. Another 10 milligrams are associated with a variety of iron-dependent enzymes, while about 3 milligrams are being transported at any one time in blood by the specialized iron-binding protein transferrin. Iron is stored in the liver, spleen and bone marrow, not as free iron, but as part of specialized iron-binding proteins, ferritin and haemosiderin. These stores are important in combatting anaemia, a deficiency of iron, and may amount to between 500 and 1000 milligrams in men and between zero and 500 milligrams in women, whose cyclical menstrual losses put them at greatest risk.

Iron metabolism operates at two levels, 'domestic' and 'imported' iron.

Most existing iron is recycled such that when a red blood cell has reached its time and 'degenerates', its package of iron-rich haemoglobin is made available once again to the bone marrow from whence it first came. Thus, the body greatly conserves iron. Imported iron is only tolerated to the level of requirements. It is well absorbed when needed and poorly absorbed when there is plenty.

Although the body conserves its iron and that iron is absorbed according to strict rules, the continual daily loss of iron from gut cells, which rapidly slough off the gut lining, means that iron must be supplied daily or thereabouts. The recommended daily intake varies between countries but is about 10 milligrams for adult men, 14 milligrams for adult women and 15 milligrams for pregnant women. These figures are arrived at by estimating normal losses from the gut, menstrual losses, bioavailability of iron from common foods used in the normal diet and the efficiency of iron absorption.

Iron deficiency anaemia is a very common problem. Between 10 and 20% of women and 2% of men suffer from iron deficiency in industrialized countries. In Third World countries, the problem is more severe and more widespread. Typically, the numbers of red blood cells are reduced and they are of a smaller size. The transferrin which carries iron in blood has very little iron attached, that is, it shows a low saturation. The capacity to carry oxygen is lowered and the person feels run down, tired and eventually breathless.

Many individuals occasionally take an iron tonic when they feel lethargic, grumpy or depressed. When taken at the recommended level, no harm will accrue and some good may come about, even if it's a placebo effect. You can achieve the same effect with diet by avoiding foods which reduce iron absorption, i.e. those rich in phytic acid (see Chapter 10), by eating foods from which iron is most available, or by increasing vitamin C intake which enhances iron absorption. Thus, for a vegetarian meal of 600–700 kilocalories, where phytic acid is likely to be high, iron absorption is 0.13 milligrams with a vitamin C intake of 7 milligrams. If, however, the vitamin C level of the meal is raised to 74 milligrams, iron absorption rises to 1 milligram. Once again, we see the sense of regularly eating a wide variety of foods and food groups.

The importance of salt in the diet has been reflected in its use as a currency and as an instrument of political power. Mahatma Gandhi and his followers marched to Dandi in 1930 to protest at the British Government's monopoly over salt production, and this event marked a turning-point in the future of India. (BBC Hulton Picture Library.)

Salt

Two minerals of vital importance to life are sodium and chloride; together they combine to make salt, which has played a central role in the history of man. Communities have developed and civilizations flourished in regions of adequate salt supply. Salt was used as a preservative either in the dry form or as a liquid brine. Salt was traded from earliest days and used as currency, hence the word 'salary'. Even in more recent times this simple molecule has played its role in political events: for example, the evacuation of the Southern Yangtse region by Red Chinese insurgents because of the Nanking government's control of salt supplies, and the remarkable pilgrimage of Mahatma Gandhi to the shores of India to protest at the British Government's monopoly of Indian salt supplies.

It is generally agreed that life began in a salty, primordial, murky pool. Sodium, as a positively charged particle, is found in highest concentrations outside cells, in the biological fluids in which they bathe. In contrast, potassium, also as a positively charged particle, is found in highest concentrations inside cells, to balance the effect of the externally located sodium. This position is maintained by an enzyme which straddles the membrane of all cells, ensuring that sodium and potassium stay in their respective quarters. Thus the cells of the body, like their earliest ancestors, are bathed in a salty environment, and the salt or sodium level is maintained constant. Any alteration has very rapid consequences.

Consider an English gentleman, who, with his mad dog, walks out in the mid-day sun. He perspires heavily in the intense heat and loses water and salt, but proportionately more water. The concentration of sodium in the fluids bathing the cells rises above the desirable level. Thirst ensues. The gentleman abuses his dog, curses the colonies and adjourns, hastily, for a cold lemonade. Later that evening, he feasts on salted groundnuts and anchovies, leading to a sudden rise in sodium absorption (sodium freely crosses the gut). Once again, there is more salt than desired in the fluid bathing the cells and he becomes thirsty. This time he abuses the servant, curses the colonies, thinks of England and downs a thirst-quenching beer. Thirst, real thirst, is a most distressful experience, aptly summarized by Coleridge's hallucinating Ancient Mariner:

> And every tongue, through utter drought
> Was withered at the root;
> We could not speak, no more than if
> We had been choked with soot . . .
> With throats unslaked, with black lips baked,
> We could not laugh nor wail;
> Through utter drought all dumb we stood!
> I bit my arm, I sucked the blood,
> And cried, A sail, A sail!

Salt, once a scarce commodity, is now widely available. Indeed there are many who would argue that it is too widely available and that our salt intake is excessively high and a major contributory cause of high blood pressure or

The physical distress brought on by severe dehydration has rarely been more vividly portrayed than in this famous picture of the *Raft of the Medusa* painted in 1819 by Théodore Géricault. (Clichés Musées Nationaux, Louvre, Paris.)

hypertension. There are two types of hypertension, secondary and essential. Only a minority of hypertensives suffer secondary hypertension, such as in severe kidney disease or hormonal imbalance. The vast majority suffer from essential hypertension, i.e. no obvious underlying cause. The general consensus is that there exists a genetic predisposition to hypertension. The precipitating factors are environmental and include both stress and diet. Foremost among culpable dietary factors is salt and the sodium atom in particular. The average intake of salt in developed countries is 10-14 grams per day and most experts argue that this is excessive. It is certainly well above physiological needs. About 40% of salt intake comes from table salt, the rest from ordinary foods (40% from unprocessed foods such as bread and milk and 20% from processed foods).

Fluorine

Another element related chemically to chlorine, the silent partner of the salt molecule, is fluorine. It is not an essential nutrient in so far as no biological function has been found for fluorine. However, recent US recommended dietary allowances of nutrients have listed a value for fluorine. It has become of importance because of its role in the prevention of tooth decay.

The link between fluoride and tooth structure was apparent back in the early part of this century in Colorado. Fred McKay, a dentist, noticed that many of his patients showed chalky patches on their teeth which, in due course, turned an unsightly mottled yellow or brown. Two facts emerged. First, this mottling process was associated with certain sources of drinking water and, secondly, this mottling did not 'seem to increase the susceptibility of the teeth to decay'. Subsequently, in England, it was found that dental caries occurred in 8% of permanent teeth in the town of Maldon, Essex, compared with a national average of 13%. The links between fluoride intake and mottled enamel and between fluoride intake and reduced dental caries soon emerged. By 1933, US health statistics showed a striking relationship between the level of fluoride in drinking water and the average number of decayed, missing or filled teeth (DMF teeth). At 8 parts per million of fluoride in drinking water, the average number of DMF teeth was eight; at 0.5 it was five; at one it was three; and at between 2 and 3 parts per million, only two DMF teeth were recorded on average.

It is now common practice for some water authorities to add fluoride to drinking water to a level of one part per million to reduce dental decay in growing children. It has been an emotive issue centring on the right to choose non-fluoridated water for those who want it, for whatever reason.

Iodine

Iodine is an essential component of the hormones secreted by the thyroid gland. When the levels in the blood are low, the thyroid swells to compensate and forms the characteristic swollen neck or 'goitre'. Many areas of the globe lie along a 'goitre belt' where low soil iodine levels lead to low plant iodine levels. If such an area is some distance from the sea, with a consequent low supply of iodine-rich seafood, and if the infrastructure of the food industry is too primitive to permit the production of iodized salt, goitre will be a common deficiency disease. In such areas, the consumption of foods which contain substances that inhibit thyroxine synthesis (see Chapter 10), will exacerbate the issue. Seafoods or iodized salt should relieve the problem, but where this is not possible iodine can be injected, in a suitable form, into the sufferers.

Zinc

Zinc is onc of several trace elements that function, almost exclusively, as co-factors in enzymatic reactions. Zinc is also involved in the immune system, although its exact role is as yet unclear. Prolonged deficiency leads to growth retardation and sexual immaturity in adolescents. Deficiency does occur in countries such as Iran and Egypt where intakes are low and where anti-nutritional factors in unleavened bread reduce its absorption (Chapter 10). A marginal zinc deficiency has been reported for the elderly, based on biochemical evidence such as low blood (serum) zinc. However, recent data suggest that biological evidence, i.e. poor wound healing, can also be indicative of zinc deficiency in the elderly. Premature babies, alcoholics and the critically ill can also encounter zinc deficiency.

Other minerals

Many other elements are found in the body. These include manganese copper, selenium, cobalt, vanadium, silicon, tin and chromium. While deficiencies can be induced in experimental animals, and then only with great difficulty, deficiencies in man are rare and really not worth substantial discussion outside of academic text-books.

Summary

- Although popularly believed to be pretty inert, the skeleton is constantly being renewed, a process which permits blood calcium to be regulated finely by the hormones, parathyroid hormone and calcitonin, and by Vitamin D.

- Iron is one mineral which frequently falls into sufficiently short supply as to produce symptoms like tiredness or breathlessness. It is absorbed, transported and utilized for oxygen transport by way of special proteins.
- The body goes to great lengths to conserve its iron and absorbs only that which it exactly needs.
- Sodium and potassium are maintained in high concentrations on either side of the cell membrane, sodium on the exterior and potassium on the interior.
- Goitre, a disease of iodine deficiency, is common in many Third World countries and can be primary, through poor diet, or secondary, through antinutritional factors in foods.
- In general the other elements do not pose problems of nutritional deficiency, except for the acutely ill or for alcoholics.

7 Vitamins

Each year, some billions of pounds are spent on across-the-counter sales of vitamin supplements in supermarkets, chemists and health food shops. Each year, millions of pounds are spent on advertising accounts centring on the vitamin content of fruit juices, breakfast cereals or what-have-you. It is common among nutritionists to scoff at this obsession we have with vitamin supplements. We can afford to scoff. We know better and yet we helped, one way or another, to create this market.

Vitamin deficiencies are, by and large, common in underdeveloped countries but comparatively rare in developed countries. Yet they do receive attention in community medicine. The incidence of vitamin deficiencies depends really on how you choose to define deficient. If it means being so deficient in, say, thiamine, so as to develop beri-beri, then deficiencies are rare. But if by deficiency you mean an observed dietary intake below that recommended, or a clinical blood analysis outside the normal range, then vitamin deficiencies are not uncommon.

The average person eating a mixed average diet is unlikely to be declared deficient by any of these criteria. The people who are likely to encounter vitamin deficiencies are the sick and the elderly. Cancer patients who cannot eat, alcoholic patients with clapped-out livers, Crohn's disease patients with malabsorbing guts, anorectics and the like are the groups most at risk. But at risk of what? Such people are very sick and vitamin deficiency is only of secondary importance to the primary medical condition. The housebound elderly might be deemed at risk on the basis of dietary investigation or blood analysis and for them a multi-vitamin supplement might be useful. Slimmers who choose to fast or adopt a very low-Calorie diet may also benefit from vitamin supplements. But I repeat that the average person eating a mixed average diet is unlikely to encounter vitamin deficiency.

In recent years, however, a new interpretation of deficiency has begun to emerge. Preventing frank repulsive symptoms is the basic definition. Preventing suboptimal intakes or suboptimal blood values is the second definition. The third one argues that optimal or slightly more than optimal intakes of certain vitamins may prevent, or rather reduce the risk of, certain diseases such as cancer or spina bifida. It's a tricky concept as we shall see, as each of the vitamins is tackled separately.

Vitamin A (retinol)

A deficiency of vitamin A will lead to night-blindness, impaired growth, impaired reproduction and abnormality of skin and tissues exposed to the environment, mainly the guts and lungs. Each year hundreds of thousands of children go permanently blind because of vitamin A deficiency. A deficiency of vitamin A is uncommon in developed countries and yet it is a very popular and topical vitamin.

There are two categories of vitamin A in the diet: vitamin A itself, found in liver, milk and dairy produce, it being a fat-soluble vitamin; and beta-carotene, found in carrots and green vegetables. The chemical term for vitamin A is retinol. Beta-carotene yields about two molecules of retinol when acted on in the gut, although not all the beta-carotene is so split. Some crosses the gut intact. The great majority of our dietary vitamin A comes, in fact, from beta-carotene.

A major reason for the widespread interest in this vitamin is the anecdotal, epidemiological and experimental evidence linking this vitamin to both the prevention and treatment of cancer.

There are many cancer self-help groups who adopt a softly softly approach to therapy, involving meditation, diet and exercise. In contrast to the high-tech approach of modern medicine, i.e. lasers, body-scanners and near-fatal drugs, this approach is a breath of fresh air, albeit for some, a last breath, drawn in dignity. Much of the dietary advice centres on fresh fruit and vegetables, emphasizing the nutritional value of vitamins A and C and dietary fibre. The basis for the involvement of vitamin A is partly epidemiological, partly experimental.

Almost two dozen epidemiological studies have sought to examine the association between cancer and vitamin A. Some have involved interviewing cancer patients and control patients as to their life-long dietary habits. The majority of such studies have suggested that an above-average intake of vitamin A is protective. Others have relied on blood retinol analysis, again showing in general that cancer patients tend to have lower than average blood retinol levels. (The most recent epidemiological investigation has not, however, found evidence to support these beliefs, although the authors of the report do call for further research – it is, after all, a 'grantable' research topic!)

There are, however, serious assumptions and grave misunderstandings in this subject, the most important being that the amount of vitamin A we eat determines our blood vitamin A (retinol) levels. It doesn't. If you encounter a population with a serious dietary deficiency of vitamin A, then you will find their levels of blood retinol are well below normal. A short period of vitamin supplementation will restore these levels to the normal range. But what of a population within the normal range? What happens to their blood retinol levels on dietary supplementation? Simply nothing.

Vitamin A, as retinol, is carried around the blood from its stores in the liver by a specialized protein, the retinol-binding protein. The limiting factor to blood retinol levels is not retinol intake but the supply of the retinol-binding protein. You can underuse its carrying capacity, as happens in deficiency, but you cannot exceed its regulation load. So what happens if you load yourself with vitamin A pills? Your liver stores of retinol increase. Now, if the rate of increase exceeds the breakdown process in the liver, and exceeds the export from the liver on the retinol-binding protein, there exists the danger of overflow into the blood of *free, unbound* retinol which has toxic properties. (Hence the good sense in carrying it on a specialized vehicle.) People have died from vitamin A overload. In fairness though, people who eat diets rich in beta-carotene (the vitamin precursor), fruit and vegetables, are utterly unlikely to suffer difficulties in blood vitamin A levels.

So, if beta-carotene-rich foods are protective, it may be due to beta-carotene *per se* and not to the derived vitamin A. If blood retinol levels are reduced in cancer patients, it may be because the retinol-binding protein is low, not because vitamin A intake is low.

Much research is clearly needed. In the world of pharmacology, retinol* and its anti-cancer properties are being studied avidly, for experiments have shown that derivatives of retinol can reduce the incidence of experimentally contrived cancer in animals, under laboratory conditions. Perhaps the miracle drug will yet materialize. Be advised. Buy shares in the drug industry.

Vitamin D

A survey of the vitamin intakes of a group of normal individuals will frequently reveal a below-average intake of vitamin D. A deficiency of vitamin D will lead to rickets and yet rickets is uncommon nowadays in developed countries. The explanation is simple. Vitamin D is the sunshine vitamin. Ultraviolet rays act on a substance present in the surface of the skin to produce vitamin D. So, theoretically, vitamin D deficiency should not be a serious problem and vitamin D supplements should be a gross waste of money. However, in the UK clinicians have become increasingly aware of vitamin D deficiency almost exclusively in one group, Asian immigrants. The reasons are complex and poorly understood.

Surveys have shown that vitamin D deficiency occurs in 24% of Glaswegian Asians who are otherwise healthy and who appear to eat an otherwise adequate diet. It may be that they have a higher genetic requirement for vitamin D or that their vegetarian habits, including the consumption of unleavened flour as chapatis, is the underlying problem. Phytic acid (see Chapter 10), which is present in cereals, can bind calcium

* Strictly speaking the drug industry is interested in derivatives of retinoic acid and not retinol, although these are closely related.

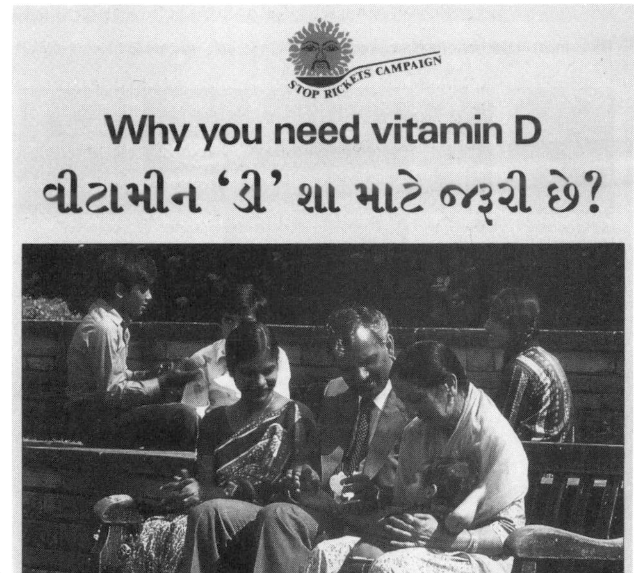

Vitamin D deficiency is particularly prevalent in Britain's Asian population and seems to be the result of a whole range of contributory factors. The answer to the problem, however, is quite simple and educational material aimed at the high-risk group provides basic advice on dietary ways of overcoming this problem. (Reproduced with permission of the Health Education Council.)

and may contribute indirectly to the skeletal disorders of vitamin D deficiency, i.e. rickets and osteomalacia. However, decreased exposure to sunlight is also likely to precipitate rickets. In one expert's opinion: 'The balance of evidence indicates that simple lack of sunshine is the main cause of the vitamin D deficiency'.

About 1% of the sun's energy is in the ultraviolet waveband and the intensity of sunlight varies with atmospheric pollution, latitude and season. The intensities in Newcastle and New Delhi obviously differ. A sensitive issue is that of the impact of ultraviolet irradiation on vitamin D synthesis in black and white peoples. Normally, 44% of ultraviolet light is transmitted through to the active part of the skin in Whites. In Blacks, this value is only 3%. This is sensitive because the lunatic fringe of certain political parties would argue that nature didn't intend Blacks to live at these latitudes and hence repatriation would be a most humanitarian gesture. However, rickets and osteomalacia are not problems among immigrant Black West Indians. The Asians, particularly Asian girls, are more protected, more housebound and suffer the further exclusion of sunlight by way of their national dress. Hence a combination of climatic, cultural, dietary, genetic and metabolic effects may act together to precipitate this problem. The solution is painfully simple: supplement chapati flour with vitamin D. It is required of margarine manufacturers to add vitamins A and D to their product, so a precedent does exist.

Vitamin E

Uncle Oswald, about whom Roald Dahl has written, stumbled across the remarkable aphrodisiac properties of an extract of the African dung beetle. Sadly, the millions of American males who gulp vitamin E tablets have never read Roald Dahl: They'd be better off experimenting with dung beetles.

Atrophy of the testicles is one of the symptoms of vitamin E deficiency in laboratory animals. It leads to sterility and is readily treated with vitamin E supplements. Therein lies the basis of the American stud's dream. If a deficiency causes testicular wasting, supplements should have the opposite effect. Of course this is nonsense, although I for one will happily participate in a well-conducted double-blind trial!

With the exception of premature infants, vitamin E deficiency is unheard of in man. Indeed the UK does not even issue a minimum or recommended daily allowance, although the US government does. One of the main functions of vitamin E in the body is to maintain the stability of cell membranes, to prevent oxidation or rancidity of the component polyunsaturated fatty acids. Thus it may be that increased intake of fats rich in polyunsaturates may require an above-average (whatever that is) intake. If it does, there is still no problem because those foods rich in polyunsaturated fatty acids are also rich in vitamin E. So, if the health-food shop representative asks you about vitamin E supplements, tell him you're into dung beetles. He'll probably stock them the following week.

Vitamin C (ascorbic acid)

In 1753, James Lind a naval physician, discovered the antiscurvy properties of citrus fruits. It was one of the earliest breakthroughs in preventive nutrition. Hitherto, seamen risked scurvy on their long sea voyages. Today scurvy is rare and yet vitamin C supplements are the biggest selling products of the vitamin industry.

Vitamin C, or more correctly ascorbic acid, is required in the body for the synthesis of connective tissue, the material which holds cells together and keeps organs apart. It has other functions but that in connective tissue seems the most important. A deficiency causes scurvy, characterized by raw-red inflamed gums in the first instance and impaired wound healing in the later stages. The requirement for vitamin C is in the region of 45 milligrams per day, which should not normally present a problem. Unless, of course, you're gullible and accept *carte blanche* the hypothesis that high doses of this vitamin prevent the common cold. This hypothesis was first mooted by Linus C. Pauling, double Nobel Laureate.

The megavitamin theory argues that 1000 milligrams (22 times the recommended daily allowance) of ascorbic acid per day will reduce the incidence and relieve the symptoms of the common cold. Such a lowly ailment can be made to appear a veritable plague, for some Canadian statisticians have estimated that we each suffer from three colds per year, which in Canada realizes Can.$270 million of lost earnings! Be that as it may, the evidence is still against any value of megadoses of vitamin C against the common cold. A review of the thirty-odd studies shows a 50:50 divide either side. A review of only those studies properly conducted to exacting scientific standards and published in reputable academic journals, subject to peer review, produces a different answer: a resounding 'NO'; A De Gaullean 'NON'. One group in the latter category did come to a 'Oui' conclusion one year, but this was reversed the following year. So, Voilà. At the British Common Cold Research Centre, volunteers were infected with known strains of common cold virus; no effect of vitamin C megadosage was seen for either protection or cure.

Megadoses of vitamin C may in fact be harmful if the doses are really excessive. The phenomenon of 'rebound scurvy' can occur. An example will illustrate this. A 24-year-old male took a handsome 15 grams of ascorbic acid daily for 2 weeks. That is 350 times the recommended dose. He then stopped taking them and 4 weeks later developed scurvy. He had induced a higher-than-average requirement by megadoses of vitamin C. Fortunately this was transitory. But it is embarrassing: 'Sorry I can't accept your kind invitation to dinner, I'm in bed with scurvy'.

Vitamin K

The bacteria which inhabit the large intestine provide the average person with all the vitamin K they require. Patients with cancer or with large bowel infections and taking massive doses of antibiotics might be at risk of deficiency. But there again they are at risk of a lot worse.

The B vitamins

There are several B vitamins: Thiamin, riboflavin, niacin, folic acid, pyridoxine, nicotinic acid, biotin, cyanocobalamin and pantothenic acid. Each functions as a co-enzyme, that is, each helps in one or more enzyme-regulated biological processes. A deficiency of any one will alter the function of its dependent enzyme with undesirable metabolic consequences.

In general, if you eat a balanced diet you will not encounter deficiencies of B vitamins. Those who advertise breakfast cereals draw attention to the thiamin, riboflavin and niacin contents of their products and point out that an average helping will provide all of a child's daily requirements for various nutrients. Yet, the fact that a fortified breakfast cereal can provide all of a child's requirements for a vitamin means little. If the western child never saw a breakfast cereal, it is unlikely that he or she would suffer B vitamin deficiencies.

Many individuals succumb to the advertisers' *spiel* about the 'pep' properties of vitamin supplements. Although it is nonsense, many people feel better for taking them. If they work, they do so on psychological grounds only. I mentioned previously (Chapter 2) that once a requirement has been met, that's it. More won't help. More petrol in a car won't improve performance. The only time that an average healthy person might benefit from vitamin supplements, especially B vitamin supplements, is when they are slimming and eating around half of the food required normally, say 800 kilocalories a day. Then, supplements may be beneficial.

In spite of the rarity of B vitamin deficiency, there are two situations which clinicians are rightly concerned about: one is drug-induced deficiencies; the other is pregnancy. Certain drugs antagonize normal vitamin function. For example, anticonvulsant drugs used in epilepsy can lead to a form of anaemia, brought about by a deficiency of folic acid. The contraceptive pill can lead to increased requirements for vitamin B_6 (pyridoxine). But these possibilities shouldn't distort the general rule that a proper diet is the best insurance against deficiency. In pregnancy, the increase in cell division leads to an increase in folic acid requirements and it is common in many antenatal clinics to prescribe folate supplements. Indeed, there has been a major interest in recent years in the role of a marginal folic acid deficiency in inducing birth defects, spina bifida in particular. Much work will have to be done to verify this claim but, in the meantime, expectant mothers should seek folate tablets, just in case.

So you see its boringly simple. Eat a balanced and varied diet and you won't suffer from a vitamin deficiency. If vitamin pills make you feel better, that's fine. You could do worse. There are of course many adverts promoting compounds which are claimed to be vitamins. These I will deal with elsewhere in this book.

Summary

- Vitamin deficiencies are common in the underdeveloped countries and in the chronically ill in developed economies.
- In general, individuals eating a normal diet do not need vitamin supplements.
- Vitamin A is required for growth, night vision and the integrity of membranes. It is derived from animal fats in its active form, retinol. Plants provide vitamin A as its precursor, beta-carotene.
- Vitamin A is transported in blood on a special protein. Consequently, protein deficiency frequently leads to vitamin A deficiency and blindness. In the West, inadequate intakes of vitamin A are tentatively linked with cancer.
- Vitamin D is the sunshine vitamin and thus vitamin D deficiency is

usually associated with certain ethnic groups in developed countries who, habitually, are extensively clothed and tend to remain indoors. Vitamin D deficiency may lead to rickets.

- Vitamin C (ascorbic acid) is required for the integrity of connective tissue. It is most noted as a possible means of preventing the common cold when taken in doses well above nutritional requirements, but the evidence is scant.
- The B vitamins are frequently added to foods, especially breakfast cereals, but it is unlikely that they limit the nutritional value of most ordinary diets. Folic acid requirements are increased in pregnancy and inadequate intakes in certain social classes are linked with neural tube defects.

8
Food allergy

In 1982 a leading editorial in the prestigious medical journal, *The Lancet*, began thus: 'Nobody doubts that food allergy exists but the subject has acquired a dubious reputation. One reason is that the diagnosis has been overworked to explain a great array of poorly understood clinical symptoms. Extravagant therapeutic claims have excited an antagonistic response, putting the subject under a cloud of controversy (which perhaps shelters a little quackery).'

This suspicion of food allergy is widespread and arises because much of the evidence which is popularly proclaimed to support food allergy has been contradictory, anecdotal, confusing and, most disturbingly, impossible to accommodate within existing hypotheses which have been tried and tested and thus far not found wanting. Food allergy is at times the butt of derision because it has, sadly, attracted its unfair share of 'nuts' who haunt nutritionally related medical matters. Well-meaning as they are, they revel in untestable anecdotal evidence with as much right to a niche in science as fortune-telling. Sometimes their predictions are correct and when they're not, well, 'you see, and it's very simple really, you see . . . [insert here any hypothesis you like to choose to explain the anomaly]'.

The term allergy was first coined by Von Pirquet in 1909 to describe an alteration in the body's reaction to foreign physical or chemical stimuli. This is a broad definition which in clinical medicine has been revised to refer solely to abnormal reactions of the immune system to foreign substances. Other abnormal reactions to foreign substances occur which do not involve the immune system. These, together with allergy, are best described as 'food intolerance'.

People are intolerant of food for a variety of reasons. Indeed, studies have shown that up to one-third of the adult population avoid one or more foods for one reason or another. We must exclude from food intolerance, the avoidance of foods for religious, ethnic or cultural reasons, and we must exclude from adverse physiological reactions the adverse gastronomic response because of the taste, texture, colour or smell of a food. The adverse physiological reactions which properly constitute food intolerance are of four types.

The first of these is the idiosyncratic intolerance arising from a hereditary abnormality of metabolism. One of these, very common in certain regions of the globe, is alactasia – an inherited deficiency in the gut of the enzyme lactase, which digests lactose, the sugar of milk. An individual who develops cramps and diarrhoea on eating a large quantity of milk may be manifesting this enzyme deficiency; while this might be mistaken for allergy it is not, since that diagnosis requires the demonstration of an abnormal function of the immune system. There are other idiosyncratic food intolerances, one such being phenylketonuria. This arises from an abnormal metabolism of an amino acid, phenylalanine. It arises in 1 in 10 000 births and can lead to mental retardation if undiagnosed. Treatment involves putting the phenylketonuric infant on a diet low in phenylalanine. Once again, the immune system is nowhere involved.

The second category of adverse food reactions is the pharmacological reactions. Pharmacology is that branch of science which deals with drugs, their mode of action and their therapeutic potential. Tachycardia, a transient quickening of heart rate, which can develop after drinking coffee, is directly linked to caffeine intake. Caffeine exerts this drug-like effect independent of the immune system. Other foods, such as those containing histamine, can induce allergy-like reactions which manifest themselves in rashes. Foods such as cheese and red wine are rich in tyramine which can help to lead to migraine by increasing the concentration of certain compounds which affect the nervous system. Similarly, chocolate contains beta-phenylthylamine which also precipitates migraine. These compounds act as drugs and do so only in those susceptible to migraine.

The third category might not seem so obvious but as the story of allergy unfolds, its importance cannot be denied. This category is best described as adverse psychoreactions. An example might best illustrate this category. From the age of 7, a young girl had encountered severe abdominal pains which deteriorated such that by 14 years of age she encountered loss of consciousness for up to an hour. Exhaustive investigation revealed the presence in her blood of antibodies to a protein in bread, gluten. We shall see shortly that this is not abnormal; indeed it is the norm to find some antibodies in blood to most dietary proteins. She was given gluten, the bread protein, under disguise without any ill effect. She was then given three small slices of bread openly: 'Within fifteen minutes she complained of abdominal pain, began to shout with pain after thirty minutes, and at that juncture appeared to lose consciousness (though she was rousable by painful stimuli). After a further ten minutes she awoke and was emotionally labile. During the whole episode there was no electro-encephalographic [testing brain function] or electro-cardiographic [testing heart function] abnormality. A psychiatrist found her to have considerable emotional disturbance and concluded that her clinical manifestations were entirely hysterical.'

Now it would be a total travesty of the truth to suggest that this finding is typical of all so-called or unconfirmed 'allergy'. However, the fact that it exists, and there are very many other citable cases, should make us wary of simply taking the patient's word as gospel. Allergy needs to be investigated. It is the fourth and most important of the classes of food intolerance and it arises because of the detrimental reaction of the immune system to one or more molecular components of the diet. Since allergy is not universal, then clearly those suffering from allergy are exhibiting a breakdown of what should be the uneventful handling of the constituents of the diet. Since it is the purpose of this book to provide a scientific basis to an understanding of nutrition, it is necessary to pause and consider the immune system and its role in allergy.

Recognition by the immune system

The primary function of the immune system is to permit each species and indeed each individual member of each species to recognize his or herself. Insert into an individual something not of its own genetic design and manufacture and it is very quickly, very systematically and very ruthlessly rejected. This causes problems for transplant surgeons. You cannot put any old heart into a candidate for heart transplant. First, it must be a human heart and, second, the donor and recipient must be immunologically compatible. Even then, there is a long period of drug prescription to suppress the immune system temporarily and a great degree of hope that the implanted tissue will 'take'. This recognition is not confined to the organs such as the heart, kidneys and bone-marrow. Individual protein molecules have been designed by each species for their own ends. Hence albumin, the main protein in human blood, has been designed by man for man's use. The dog has designed its own albumin; so have the cat and the cockerel, and all other mammals. All of these albumins are roughly similar, serving the same function in all species. But each species has its own quirk of design and hence each species recognizes its own albumin and rejects albumin from other species. Clearly each species recognizes each of its own million or so individual proteins. For, although a glass of red wine is a glass of red wine, between all glasses of red wine there is great variation in taste, bouquet and all the other ethereal attributes, including potential for inducing hangover!

It is important to note that as we move downward in biological complexity from organs, such as the heart, to the less complex individual protein molecule, then within-species rejection is less of a problem. I can be injected with albumin from any other human with no adverse effect. I cannot be injected with albumin from another species, or if I am, I will mount an immunological response. Yet I cannot receive a human organ implant without exhaustive cross-checking of compatibility and only under rare circumstances can I accept an organ from another species. In essence, each species recognizes its own molecular design and each individual recognizes, to a considerable extent, his or her own tissue and organ design.

It is equally important to note that the immune system pays no attention to interspecies differences in dietary fat and carbohydrate design, since these compounds are more or less identical from piranhas to presidents and cockroaches to queens. Hence, there is no need to recognize them – they're all the same.

The immune response

The immune system itself operates on two levels, the cellular and the molecular. The cells are known as lymphocytes and the molecules as antibodies. The cells manufacture and secrete the antibodies; but since that aspect of the immune system which will concern us is the molecular aspect we will concentrate on the antibodies, with the occasional reference to lymphocytes.* Antibodies are proteins. There are five classes of antibodies found in the blood, each identified by a letter, G, A, M, E and D. Antibodies are, chemically speaking immunoglobulins (Ig), hence IgG, IgA, IgM, IgE and IgD. The most important of these in the blood is IgG. There are millions of antibodies of the IgG class in blood, each directed against a specific foreign compound. For an influenza virus, there is a specific antibody; for polio, there is another. Each can only do its own job. Thus each foreign agent in the body has its own special gang of pursuing antibodies. These antibodies protect us by marshalling their particular target into specialized cells to be destroyed. They are like secret policemen (antibodies) taking anti-State activists (foreign proteins which have entered the body) to prison (special cells in the blood and liver) for liquidation.

The uptake of foreign bodies

It is easy to envisage this type of antibody production when we present foreign bodies with easy access to tissues such as at the site of wounds or when we are desperate enough to transplant tissues. It is not too easy to

* Lymphocytes are found in the blood and, in large quantities, in lymph nodes located in specific regions in the body. When you have a sore throat, your lymph nodes about the throat swell in activity and can be readily felt. Similarly, a cut and infected finger may well give rise to a lump in the armpit, indicating another swollen lymph node.

envisage how this internal defence system might be directed against the constituents of a normal meal. I have previously stated that the immune system will only respond to foreign proteins and will do little, or indeed nothing, about dietary carbohydrates or fats which are structurally identical for all species. The problems which might be created through an immune response to diet are therefore primarily confined to dietary protein. In Chapter 2, however, it was clearly stated that dietary proteins are normally broken down in the gut to their constituent amino acids by specialist digestive enzymes. These amino acids are then absorbed and taken to the various tissues for synthesis into our own proteins, according to our own genetic specifications. It is this way that pork or pea or prawn protein can be transformed into human protein for human functions according to the human genetic code. So how, if at all, do intact proteins cross the gut wall to stimulate an immune response? The answer is 'with great difficulty', but *not* 'never'.

To begin with, there exists a powerful line of defence down at gut level, specifically designed to prevent this possibility. This would have to be overcome and is overcome, hourly, but by so few molecules that unless one specifically set out to look at this possibility in an experiment, one would never observe this tiny invasion. Nowadays, with powerful electron microscopes and analytical techniques capable of detecting as little as one-thousand-millionth of a gram of protein, this hourly trickle of proteins becomes evident. The moment such a foreign protein overcomes the first barrier at gut level, the circulating antibody system disposes of it as described above.

At this juncture one might summarize, parable-wise: A 'sleeking' cowrin' timrous beastie, i.e. a mouse, ambles about a wheatfield, gorging himself on grain and grub. He has antibodies, both in the circulation and at the gut wall, against the proteins of both grain and grub. One day he boldly enters the farmhouse and sees before him a plate of scrambled egg which he sniffs and, satisfied as to its olfactory properties, begins to nibble. Because he has never eaten egg before he cannot have antibodies against egg either in the circulation or at gut level. He returns to the field to sleep off his feast. As he does, a few egg protein molecules cross his gut with relative ease. No barriers as yet exist. The moment these proteins enter the circulation, shock waves resound throughout the immune system, sirens wail and an antibody to egg is produced, slowly at first and then in quantity. Some egg-antibodies are produced by the gut and some in circulation. The offending egg protein is dispatched for annihilation. The mouse wakens, recalls his banquet and returns to the dining hall. This time the gut is ready. Now very very little, if any, egg protein will cross the first line of defence but if some should, the second line of defence, the circulating antibodies, will deal with the intruder. This process is physiological – that is, it is normal. Consequently, one expects to find antibodies in our blood to all proteins which we have ever eaten. Not surprisingly, those investigating the neurotic young girl who 'reacted' to bread, found antibodies to bread protein in her circulation. I have them; so have you if you've ever eaten bread.

Food allergy

Mechanisms

True food allergy involves an adverse physiological reaction to a food which is meditated by a defective immunological response. In general, allergy (or hypersensitivity) can be classified into 'immediate' and 'delayed' types. The immediate reactions generally involve antibodies of the IgE class, while delayed reactions usually involve those of the IgG class. The symptoms which can be provoked in food allergy are wide-ranging and I shall classify them into 'objective' and 'subjective' symptoms. Vomiting and diarrhoea are objectively verifiable symptoms of food-linked allergy. Cramps or aches are subjective symptoms of food-linked allergy. The objective symptoms are more amenable to the rigours of experimentation and clinical investigation, and hence it is with these that most progress has been made. They include: sudden death from anaphylaxis (rapid constriction of air passages in the lung); vomiting; diarrhoea; anaemia; skin disorders such as urticaria (hives) and eczema; respiratory problems such as asthma and hay fever (rhinitis); and possibly behavioural disorders such as migraine. The symptoms which I classify as subjective read like those of a mammoth hangover: irritability; depression; lethargy; loss of memory; loss of libido; aches; cramps; blurred vision *ad nauseum*, literally!

The foods which lead to allergic reactions are legion. However, the most culpable seem to be cow's milk, eggs, nuts, fish, especially shellfish, cereals, chocolate, citrus fruits and tomatoes. Tomatoes, citrus fruits and strawberries tend to be most associated with skin-related allergies, while cow's milk seems strongly linked with gastrointestinal allergic reactions. Artificial colouring agents are also linked with food allergy or at least with food intolerance, for it is not clear how such non-protein molecules, and quite small molecules at that, can be involved in an immunological derangement. However, small non-protein molecules can produce allergic reactions by adhering to one of our own proteins in the blood or gut and in so doing render the composite structure 'foreign', leading ultimately to its disposal. Consider penicillin, to which many people are hypersensitive.

The precise mechanisms of food allergy are poorly understood. In delayed allergy, it is possible that the molecular lattices formed from, say, anti-egg antibodies (of the IgG class) bound to the specific foreign protein, in this case egg protein, persist in circulation for longer and in higher proportions than is desirable. Whether this arises because of excessive uptake of the offending protein, which itself could result from a defect in the gut's first line of defence, or from defective clearance of the lattice, or both, is not known. In the immediate type of allergy, it is likely that a cell known as a mast cell is involved. This cell is found in tissues such as the gut, lungs and skin. On exposure to a foreign protein, the mast cell becomes 'primed', i.e. coated with antibodies, say anti-egg antibodies, this time of the IgE class. Upon further exposure, egg protein reacts with the appropriately 'primed' mast cell and causes the cell to release a variety of pharmacologically active (drug-like) compounds designed to cause an inflammatory response. In allergy the extent of this response is excessive, leading to nasty symptoms such as anaphylaxis, potentially fatal. For example, histamine is released by such cells. If excessive histamine is produced, painful consequences can occur, hence the frequent use of antihistamines.

Diagnosis and treatment

A person suspected of having diabetes is subjected to a standard test known as the glucose tolerance test. They receive a standard oral dose of glucose and their blood glucose is monitored half-hourly for 2 hours. Ideally, the level should rise rapidly and fall rapidly. In diabetes it rises to an unusually high level and remains high for a long period. The test is simple, standard and reliable. No such test exists in diagnosing food allergy. Sampling blood for measurement of antibodies to suspect foods or for complex lattices of antibodies bound to their appropriate foreign protein are hopelessly unreliable. You get positive results when they should be negative, and vice versa.

A popular approach to diagnosing suspected food allergy is to put the patient on an 'exclusion' diet which varies from clinic to clinic but basically sets out to avoid the commonly offending foods. Possible food allergens are introduced one at a time at 3–4-day intervals. If no symptoms occur upon introduction of one food, the patient proceeds to the next until an offending food is identified. At that point the patient returns to the previous satisfactory diet and proceeds onwards and so forth. The procedure is protracted, non-technological and only partly effective, because its limitation lies in not excluding the power of suggestion. The only sure way of excluding the power of suggestion is to identify potentially offending foods by the method just described and then to conduct a double-blind food challenge. In this method the patient is offered a selection of foods either disguised in a soup or purée, or in capsule form, such that neither the patient nor the attending physician knows which test contains which food. Only in that way can patient and observer bias be ruled out. Half of the food challenges contain foods which were perfectly well tolerated in the preliminary process of the exclusion diet. The other half of the food challenges contain the alleged problem foods. A patient with genuine allergy will 'react' only with the introduction of the offending food. Those who have a pyschological problem will react randomly and the chance of their reactions coinciding with the consumption of the offending food is remote. It is thus the only acid test of food allergy.

To illustrate the need for the acid test of double-blind challenge, let me describe the results of a study carried out by one of the USA's leading allergy specialists on children whose *parents* or *family doctor* had diagnosed food allergy without a double-blind food challenge: 'The data which follow were collected during food challenges of 171 children, 119 of whom were 3 years of age or older and 52 were younger than 3 years of age. The sole criterion for admission to the study was that the child presented with an impressive history of an adverse reaction associated with ingestion of one or more foods.' In the course of a double-blind challenge using suspect food or placebo capsules, '... symptoms were evoked in 49 (41%) of the 119 children of 3 years or older ... In the 52 children under 3 years, symptoms were evoked in 22 (42%) ... Thus in about 60% of patients tested the adverse reaction could not be confirmed and when the foods under suspicion were re-introduced into the diets of the children, they were tolerated without difficulty'. The symptoms involved were in the gastrointestinal tract, skin and respiratory system. Children of that age cannot invent symptoms of that nature, so 60% were misdiagnosed by overenthusiastic parents and general practitioners.

It is this enthusiasm which bedevils the study of allergy and a moment's reflection on the lunatic fringe will illustrate its dangers. There are some who claim that they can diagnose 'allergy' by an electromagnetic technique. A magnet is held by a thread in front of the patient who holds two foods, one in each hand. The magnet will tend to veer toward the potentially offending food. Believe me, I've seen it on television! Then, there is the 'science' of radionics, which uses a hair sample to determine allergy. It has been elegantly exposed. A consultant dermatologist at the Royal Infirmary, Glasgow, was treating a patient with classic allergic reactions to eggs and fish. The allergic symptom was angio-oedema, a nasty swelling, in this case of the face and eyes. The patient also had positive reactions to egg and fish using the somewhat unpredictable laboratory analytical tests of allergy. A sample of this patient's hair was sent for radionic testing which revealed allergy to 'cow's milk and cheese'! Pulse-testing is another questionable technique whereby the patient measures the increase in pulse-rate on eating foods. According to this theory the offending foods will cause pulse rate to quicken. Try it yourself! And then there is the 'autoimmune urine therapy' which recommends the consumption of one's own urine. It isn't toxic, since the octagenarian former Indian Prime Minister, Mr Moraji Desai, drank his own urine, daily, from the age of 20. There is however little evidence to suggest that such 'therapy' works.

Vigilance in exposing quackery is required in other more serious instances. The 'total allergy syndrome', whereby people are allergic to the twentieth century, has attracted considerable media attention. It is poorly defined and is largely a subjective diagnosis. One explanation is that such people engage in 2–3 minutes of hyperventilation, i.e. rapid and exaggerated breathing.

A particularly difficult problem is that of Dr Ben Feingold's hypothesis that hyperactivity in children arises from an 'allergy to' or 'hypersensitivity to' certain food additives (particularly salicylates) such as artificial colouring and flavouring agents. Many thousands of parents, distraught with an insomniac, irrepressible, ever-active child, resorted to the rigours of just what such a diet meant, natural foods only, which to an American child is like orange juice to alcoholics. Widespread uncorroborated anecdotal claims were made in every medium other than in scientific literature. In a somewhat unusual response, The Nutrition Foundation of the USA, a non-profit-making public institution, set about a multi-centre double-blind trial using special biscuits with or without the inclusion of the most notorious of Feingold's additives. They concluded: 'The studies . . . represent the efforts of dozens of investigators over more than a four year period at a cost which has probably exeeded the million-dollar level. It is our opinion that the studies already completed, proved sufficient evidence to refute the claim that artificial food colourings, artificial flavourings and salicylates produce hyperactivity and/or learning disability.'

Conclusions

How long is a piece of string? That's about the measure of the problem of food allergy is. If you wish to confine food allergy to clinically verifiable, immunologically based adverse physiological reactions, then the string is short. If you want to include all forms of adverse reactions, physiological and psychological, diagnosed by whatever means, encompassing immunological and metabolic defects, then the string is very long.

Summary

- Food intolerance may arise for four reasons: inborn errors of digestion or metabolism; pharmacological factors; psychological factors; or allergy. Only allergy is associated with an abnormal response of the immune system.
- The carbohydrates and fats in nature are more or less identical between

species. Proteins aren't. Each species has its own blueprint and any intruder is removed by antibodies produced by lymphocytes.

- Food proteins are 'foreign' and as such are excluded by two lines of defence: antibodies secreted onto the surface layer of the gut which recognize and bind foreign proteins; and a mix of antibodies in the blood which mop up any intruders.
- Allergic reactions are associated with certain tissues and certain foods. The symptoms can be delayed or immediate.
- The diagnosis of true food allergy requires repeated challenges of the patient with the suspect food and an 'innocent' food, both disguised to prevent any subjective bias.
- The incidence of true food allergy is low.

9
Coronary heart disease: the crowning controversy

Background

The Coronary Care Unit at your local general hospital represents the front line in what has been dubbed the 'battle against heart disease'. Each day, with monotonous regularity, the casualties of battle are wheeled in, plucked from their apparent pink of health, to lie drugged and frightened, surrounded on all sides by the sight and sound of modern high-tech medicine. Those who arrive do not follow any particular pattern: they are young and old; they are male and female; some smoke, some don't; some will have had symptoms, others won't; some will survive, some won't. Consider how this diffuse pattern differs from that of patients being treated for another disease equally rampant in our modern society – influenza.

What ultimately distinguishes a patient with 'flu from a non-patient is the presence of a specific virus which is identifiable and is positive proof of cause and effect. You have the bug, you get the 'flu. You don't have, you don't get. Thus are the populations with and without 'flu distinguished. But for those suffering heart disease, there is nothing to distinguish them – absolutely – from those free of the disease. Because the distinction cannot be absolute, it has to be made relative. The two groups are apparently similar. But one group tends to have relatively more or less of one or more attributes. For example, the populations of people with and without heart disease each have cholesterol in their blood but those with heart disease have, as a group, more than those free of the disease. No single extraneous force is known, the presence of which causes the disease or the treatment of which cures the disease. In the absence of such, the epidemiologists have put together a profile of the hypothetical heart attack victim, drawn from data gathered on half a billion real victims, now spanning three decades. The following facts emerge.

Age

Heart attacks are very rare in infancy and youth. The incidence increases from early 'middle age' onwards to reach its peak among the elderly.

Sex

Men are more prone to heart attack than women in the premenopausal phase. Postmenopausally, however, the incidences are equal.

Weight

People who are very heavy may be at a greater risk than those at the ideal body weight. This is not, however, a direct effect as many seem to think ('his heart will never stand the strain of all that weight . . .'). The risk arises because the obese person is more likely to develop diabetes, show high levels of blood cholesterol and get high blood pressure than are normal individuals of normal weight.

Blood pressure

Individuals with high blood pressure are 'hypertensive', in medical jargon. The greater the elevation above the norm, the greater the risk.

Blood cholesterol

Cholesterol in the blood is carried about in lipoproteins (see Chapter 2) of varying density: very low- (VLDL), low- (LDL) and high- (HDL) density lipoproteins. Elevated blood cholesterol levels increase the risk of coronary heart disease. This elevation in risk is associated only with the LDL fraction, however. HDL cholesterol is in fact protective, in that higher levels are associated with lower rates of coronary heart disease.

Smoking

There is a very simple and direct link between smoking and heart disease. The more you smoke, the greater your risk. You begin to reap the benefit of a gradually reducing risk as soon as you stop smoking.

Personality

Individuals can be classified as Type A or Type B. Type A individuals are aggressive, dominated by schedules and deadlines, and competitive. Type B people are 'laid-back'. Type A people are allegedly more prone to coronary heart disease.

Geography

The incidence of heart disease is not uniform in any country. For example, in the UK, Western Scotland has a 50% higher heart disease incidence than South East England; and within England, incidence varies from town to town. One possible explanation of geographical variation is the hardness of the local drinking water, hard water being protective. But, of course, lifestyle must play a role – life in the Mid-West of the USA isn't quite like that in Southern California.

Genetics

The Japanese have a low incidence of coronary heart disease. On emigration to Hawaii, however, the incidence increases and on further eastward movement to California there is yet another increase in incidence in Japanese people. So the protection seen in Japan is not necessarily genetic. However, there are apparent familial tendencies in coronary heart disease: one very often sees several brothers with the disease.

Exercise

All exercise, unless overdone, is good for the heart, though this is not always obvious during the exercise. Aerobic exercise – basically prolonged steady exercise which doesn't leave you breathless – helps to protect against coronary heart disease by raising the HDL cholesterol levels and by increasing the ability of the blood to disperse clots.

Diet

Of all the factors implicated in heart disease, diet has received the widest attention, with advice and counter-advice abounding. Much of the content of this chapter will deal specifically with this problem. The 'problem' of diet in coronary heart disease is dominated by the manner in which diet can influence blood cholesterol. It behoves us therefore, before pursuing this matter of diet, to consider blood cholesterol in more detail and to try to explain the role of diet in coronary heart disease.

Cholesterol – whence it comes and where it goes

Few biological chemicals have been given as much publicity as cholesterol. It is to be avoided and feared, if the popular press is any barometer of medical opinion. But this most ubiquitous of chemicals is misunderstood and misrepresented.

There are three major biological functions of cholesterol, the most important of which is as a component of biological membranes. As we noted in Chapter 2, these membranes surround every cell and every subsection of every cell. In this manner are the specific functions of each cell and each section of each cell kept separate. These membranes are made up of phospholipids, proteins and cholesterol. The proteins float about in the membrane and act as enzymes, such as those to transport amino acids into cells, and receptors, which bind to specific hormones, vitamins or prostaglandins. Cholesterol plays a key role in retaining the fluid nature which is an essential feature of membranes.

The second major function of cholesterol is as a precursor of several hormones, including corticosterone, aldosterone and the sex hormones. These chemicals have important biological roles but are active in small quantities. Hence this function, while important, does not make a great demand, quantitatively, on body cholesterol.

The final requirement for cholesterol is as a precursor of bile, or more specifically of the bile acids. Two of these are synthesized in the liver from cholesterol and are known as primary bile acids. They either enter the gut directly from the liver or via their storage area, the gall bladder. In the small intestine they solubilize dietary fat and render it susceptible to enzymatic digestion. As the bile acid – fatty acid sphere moves down the gut, the fatty acids are absorbed. Subsequently the bile acids are reabsorbed, although before reabsorption some are altered slightly by gut bacteria to form secondary bile acids. These primary and secondary bile acids return to the liver for re-secretion. The whole process is repeated up to fifty times a day in the average person. Only a small fraction is lost in faeces.

All these functions account for the 140 grams of cholesterol present in the average adult man. The next question is where does it come from?

Diet is obviously one source. Most Western adults consume about 500 milligrams per day and about 40% of this is absorbed. If cholesterol intake is increased to, say, 150 milligrams per day, the percentage absorbed falls; conversely, if one pursues a low-cholesterol diet, the percentage absorption is increased. As a result, dietary modification of cholesterol intake has far less effect on the amount actually absorbed than is commonly believed. The vast majority of people, consuming a wide range of cholesterol, will absorb 150–250 milligrams per day. The dietary sources of cholesterol are confined to animal produce, so strict vegetarians, vegans, consume no cholesterol. Plant membranes have a comparable 'sterol' which is not absorbed at all (sitosterol).

Now, if vegans do not consume cholesterol and yet cholesterol serves an essential function in the body, where else does this cholesterol come from? It is synthesized by two organs, the liver and the intestine, the former being quantitatively the more important. Strictly speaking every cell in the body can make its own cholesterol, but cells generally tend to rely on that made by the two export tissues. The synthesized cholesterol is despatched into blood as part of a lipoprotein belonging to the VLDL class. Lipoproteins, discussed in Chapter 2, are made up basically of a core of fat (triglyceride) surrounded by free cholesterol, phospholipids and proteins. In the case of the VLDL, the core is fat (triglyceride) which has been synthesized by the liver (or the gut). This particle arrives at the muscles and fat depots and deposits its inner core. The remaining particle returns to the liver whence it re-emerges, this time as a LDL. It has therefore graduated from very low- to low-density lipoprotein.

The essential difference between the VLDL and LDL is that the latter contains little or no triglyceride in its core. This core material has been replaced by cholesterol. The average compositions of VLDL and LDL are shown in Table 9.

The LDL enter the peripheral circulation. The cholesterol found in the plasma is predominantly of the LDL type. Just to complicate matters, there is the third type of lipoprotein, the high-density lipoprotein (HDL). This fraction contains a lot of protein (50%), cholesterol esters (20%) and phospholipids (26%). VLDL represents a package of fat on its way to adipose tissue or muscle; LDL represents a package of cholesterol on its way to all cells; and HDL represents a package of cholesterol from the tissues to the liver for metabolism to bile acids. Thus is summarized a score of tomes and a quarter of a century of research.

The discovery of how the LDL particle delivers its package to cells other than in the liver and gut was the most significant in this field over the last decade and won the 1985 Nobel Prize in Medicine for the Dallas-based scientists Joseph Goldstein and Michael Brown. The LDL particle does so by way of specific points of entry on the cell membranes: LDL receptors. These function like import agencies. When the LDL supply is above normal, the 'value' falls, and the number of agencies or LDL receptors falls. When the LDL supply is below normal, the 'value' rises and the number of agencies or LDL receptors rises. Either way, the amount flowing in depends on demand inside the cell. When demand is met, import closes down and home production of cholesterol ceases. When external supplies are very low, not only does the number of importing agencies (LDL receptors) increase, but home production of cholesterol also rises. The bottom line is this: the cell regulates its cholesterol level in a very strict manner by manipulating the number of receptors and the extent of home production of cholesterol by the cell itself. Under no circumstances does a normal cell allow cholesterol to accumulate. Yet under the

Table 9. *Average compositions (%) of VLDL and LDL*

	VLDL	LDL
Shell		
Phospholid	18	22
Protein	9	21
Free cholesterol	7	8
Core		
Triglyceride	54	11
Cholesterol (ester) [a]	12	37

[a]Cholesterol exists either 'free' or bound to a fatty acid as an 'ester'. The latter is even more insoluble than free cholesterol and thus hides in the safety of the lipoprotein core.

pathological conditions of coronary heart disease, this does happen. Now is the moment to consider this pathology.

The pathology of coronary heart disease

Heart disease is an all-embracing term encompassing several different pathological processes. The most fundamental pathological feature of heart disease is atherosclerosis (*athero* is a Greek word for 'gruel'; sclerosis denotes 'hardening'). It is a fancy term for hardening of the arteries. You do not expect to find extensive atherosclerosis in infants. But you probably have atherosclerosis in your coronary arteries – and probably not a symptom in the world. For atherosclerosis, whilst uncommon in infancy, is widespread in adulthood in developed countries. It is in fact part and parcel of senescence. Your hair falls out, your breasts droop, your teeth and gums rot, your libido wanes, your limbs stiffen, your bones break more easily, your memory fails, your arteries harden, and you age. It is therefore essential to establish at the outset that hardened arteries do not necessarily spell heart disease. However, the great majority, though not all, of those who succumb to heart disease do have atherosclerosis. Why some succumb to heart disease when most of us have some degree of atherosclerosis is the first of several mysteries you will soon encounter. First, an explanation of the mechanism of atherosclerosis.

An artery, like a tyre has many layers. The innermost layer, in direct contact with flowing blood, is called the endothelium. It is only one cell thick but for all its delicate structure it is a powerful and influential layer. Underneath lies a layer of the protein elastin which is why when you pull an artery (not that you're likely to have the chance except perhaps if you stumble across one on your plate – liver is a good possibility), it is like an elastic band. Remember, it needs to be to pulsate seventy times a minute as blood is pumped through. Underneath the elastin fibres lies the main body of the artery, known as the media, which is rich in muscle cells of the smooth type. These smooth muscle cells contribute strength to the artery during pulsation. The appearance of a normal artery is shown in (*a*) of the diagram on the facing page.

A hardened artery differs radically. The endothelium is not always as complete as usual. The smooth muscle cell layer is relatively unchanged, but the layer between the elastin fibres and the endothelium is swollen as shown in part (*b*) opposite. The extent of this swelling dictates the extent of reduction in blood flow which, in turn, dictates, in part at least, the onset of symptoms.

Remember that atherosclerosis is derived from the Greek words 'gruel' and 'hardening'. The gruel is fatty and very rich in cholesterol. Some of it is in cells (smooth muscle) which have migrated into the lumen of the artery. Some of it lies crystallized, embedded in a mish-mash of decaying cells. Fibrous proteins usually overlie this gruel and impart a degree of hardness. You will have encountered fibrous proteins before. Have you ever noticed how a scar lacks the soft supple elastic features of the surrounding skin? That change in texture is due to the deposition of fibrous proteins. To summarize then, a lesioned artery has a swelling, just underneath the inner surface, which is rich in cholesterol, decaying and dead cells, crystals of cholesterol, some chalky deposits and the scar-like proteins of fibrous tissue.

At this moment of time there are several equally attractive hypotheses to explain this lesion and in particular to explain the accumulation of huge deposits of cholesterol. Nothing will be gained from detailing them at this juncture.

The cholesterol which accumulates under the surface of a lesioned artery has come from blood. On that there is near unanimous agreement. This cholesterol has been derived from that carried on the LDL carrier. On

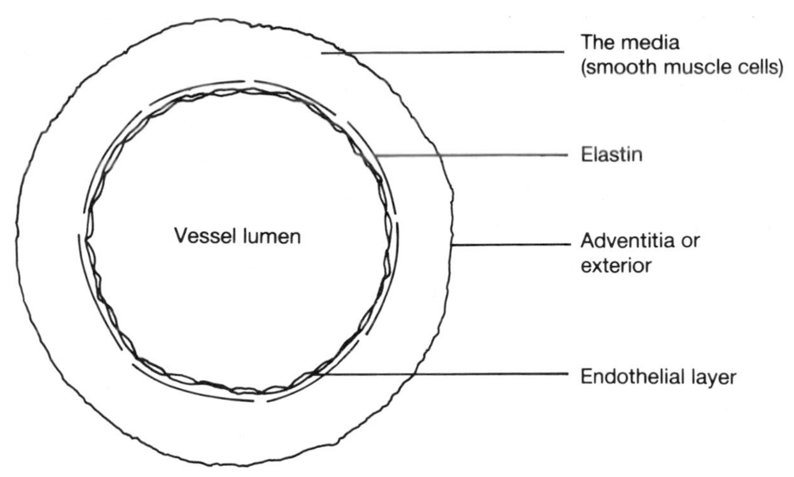

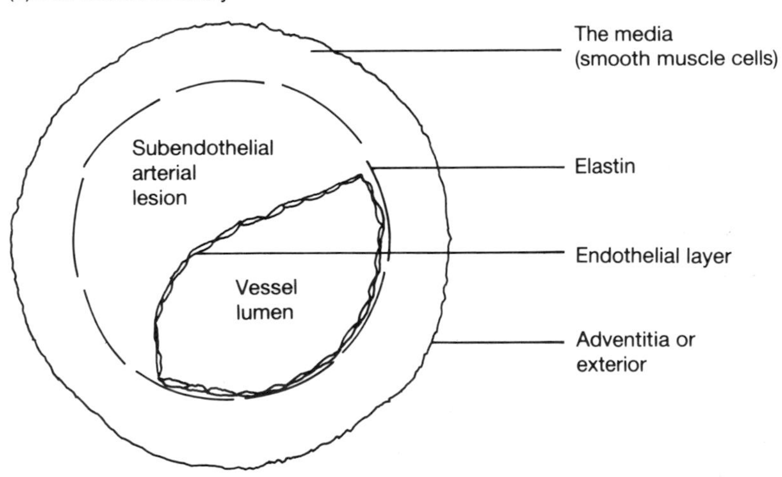

The structure of normal and half-occluded arteries.

that there is near unanimous agreement. It might seem likely, therefore, that the more cholesterol in blood (in fact, the more LDL cholesterol in blood), the greater the accumulation in arteries. Hence the link between blood cholesterol and heart disease. It is possible. But we shall see.

The most striking implication of hardened arteries is that the flow of blood becomes partly restricted. The more extensive the lesion, the greater the restriction, and the greater the likelihood of symptoms. The first possibility is the development of angina, a term used to cover a transient stabbing pain in the chest. It occurs when the flow of blood around the heart muscle, in the coronary arteries, is restricted. If the restriction is substantial, and if the heart is demanding its fair share of oxygen to hawk its owner about a golf course, it complains at an inadequate rate of supply of oxygen. It simply won't work without oxygen. The patient gets breathless and pain ensues. The patient stops and rests. The demand for oxygen drops to around that which the partly narrowed artery can supply. The pain subsides. The patient rests. Such a person learns to live with angina.

For others, heart disease becomes more dramatic. One moment they are fine; the next they are breathless tortured by a stabbing pain in the chest. They have suffered a heart attack. The process can be outlined as follows. Blood was flowing down a partly occluded artery in the coronary tree surrounding the heart muscle. For some reason, the usual ability of the endothelium, the inner lining of the artery, to prevent blood cells adhering to the sides of the vessel is lost (at the site of the lesion only). Small cells known as platelets stick to the injured endothelium, which may for example have ruptured (like a boil). More platelets stick until the platelet mass forms a clot or 'thrombus' large enough to block all blood flow. That part of the heart depending on that part of the coronary tree for blood flow is then injured. This is a myocardial infarction, a heart attack. It can be mild or massive. It can be fatal. It has arisen through two pathological processes: atherosclerosis, which partly occludes arteries; and thrombosis, which in an unpredictable manner finishes the job off at a later date.

The diet–heart disease dilemma

Let us return now to the question of diet and coronary heart disease. The level of cholesterol in our blood is determined by the balance of two factors; our own inherent capacity to make and degrade cholesterol, and the contribution which dietary factors make. What concerns many in the field of public health is that diet is making an unfair contribution which could, in part, be altered by improved nutrition.

The nutritional impact on serum cholesterol is a much-studied field. Indeed, one might fairly say that it is an over-studied field. Cholesterol in blood is easy to measure and the assay can be automated to tackle each sample in less than a minute. Thus, any moderately equipped laboratory can do cholesterol research. One paper has reviewed the literature and identified 147 different factors, many of them dietary, which alter blood cholesterol.

As I have mentioned earlier, dietary cholesterol in its own right, does not have as dramatic an effect on blood cholesterol as is commonly believed. Fat has a bigger effect. In particular, saturated fat can increase blood cholesterol and polyunsaturated fat can lower it (see Chapter 2 for a discussion of these fats). Tens of thousands of animal experiments and thousands of human studies have verified this. The effect, however, depends on two factors. First, the higher the starting value of blood cholesterol, the greater the impact of polyunsaturated fats; and the greater their intake, the greater their effect. However, before charging out to buy shares in the edible oils industry persist and read on. Protein also affects serum cholesterol levels, although the evidence for an effect in man is arguable. Animal protein elevates and vegetable protein depresses cholesterol levels in blood. Certain types of fibre reduce blood cholesterol levels, especially the fibre in legumes, fruits and oats. All these together influence serum cholesterol in so far as an excess of caloric intake over caloric expenditure leads to obesity of some degree, and blood cholesterol rises as body fat stores rise. So is it possible to lower blood cholesterol by dietary means? As I have previously pointed out, that depends on how high the initial blood cholesterol level is and on how drastic is the dietary intervention.

The Oslo Study provides a suitable illustration. Men with cholesterol levels in the high end of the range were chosen. One group received intensive nutrition counselling to alter the pattern of fat intake such that total fat and saturated fat intakes fell while polyunsaturated fat intakes increased; they also cut down on smoking. The results are given in Table 10.

This degree of dietary intervention was clearly successful in reducing

Table 10. *Effects of dietary intervention and no intervention (control group) on serum cholesterol levels and incidence of death from coronary heart disease in men in the Oslo Study*

(a) Baseline values

	Intervention group	Control group
Age (years)	45	45
Heart disease cases	Nil	Nil
Smokers (%)	70	70
Serum cholesterol (milligrams per 100 millilitres)	328	329

(b) after 5 years	Intervention group	Control group
Dietary fat (% energy intake)	28	40
Saturated fat (% energy intake)	8	18
Serum cholesterol (milligrams per 100 millilitres)	263	341
Coronary deaths per 1000	31	57

blood cholesterol levels. Taken together with a reduction in cigarette smoking and in blood pressure in the intervention group, total mortality from coronary heart disease fell dramatically (31 *versus* 57 per 1000).

So you bought shares in the edible oils industry and you see them looking good. The case for mass dietary intervention seems obvious. Read on.

There have been many mass intervention trials since this nutritional hullaballoo first surfaced after the Second World War. They have several features in common, the most important of which is their failure to answer the question which first prompted them. Take the example of the Los Angeles Veterans Administration Hospital trial.

Some 1000 men were involved in a study lasting seven years. They differed from those in the Oslo Study in that they were *not preselected from the top end of the range of blood cholesterol levels. They were average citizens with average values for blood cholesterol.* The men were assigned to either a standard hospital diet high in fat, mostly saturated, or to a low-fat diet with most fat polyunsaturated. Blood cholesterol levels behaved as predicted: those on the high-polyunsaturated-fat diet had lower levels of blood cholesterol and they had fewer heart attacks. Those shares look better by the line. Alas, in a curious twist of nature, although this group had fewer deaths from heart disease, they had more deaths from cancer. It balanced out. Death rates in the two groups were equal. A similar result was observed in an equally extensive study in Finland. Polyunsaturates lowered blood cholesterol and with it lowered deaths from heart disease. However, this decline was compensated for by an equal increase in deaths from other causes, leaving total mortality equal.

There have been many other studies which set out to tackle this problem, all of which have been equally inconclusive. Either there were no changes in total mortality, with heart disease and cancer balancing the equations of death, or there have been equal changes in *both* the intervention group and the control group during the study period. The MRFIT study (Multiple Risk Factor Intervention Trial) is an example of the latter. One group received massive counselling on smoking, high blood pressure and diet. The other group received no particular attention. Yet both groups showed reduced blood cholesterol and reduced deaths from heart disease.

In an attempt to bring some sense to this spiral of intervention theory, one leading cardiologist chose to quote Oliver Cromwell who, in 1605, addressed the Assembled Church in Scotland thus: 'Gentlemen, I beseech thee in the bowels of Christ, Think it possible you may be wrong'.

Wrong? Wrong to want to change our diet to stamp out heart disease? No, wrong to try and turn the whole population into patients eating a prescription diet in the somewhat uncertain hope that some good will accrue. Better, many argue, to propose only modest dietary changes for Mr Everyman and substantial changes for those we know to be most at risk, those at the high range of blood cholesterol. For the only successful intervention programmes, using diet or drugs or both, have preselected those with serum cholesterol levels in the top 5–10% of the distribution.

This point of view is opposed by the 'interventionists' who argue thus: people at the top end of the cholesterol scale have five times the risk of a coronary event as people with levels in the middle of the range. However, there are four times as many people in the middle than at the top of the range. So, whilst the risk for these in the middle is certainly less, they account for more cases of heart disease because there are more of them. Let us quantify this. In one study it was shown that of 10 000 subjects, 662 were individuals with cholesterol levels greater than 300 milligrams per 100 millilitres of blood. Of these 662, 62 died of coronary heart disease – an incidence of 62 per 662 or 9.1%. In the middle range (200–240 milligrams per 100 millilitres) there were 3162 men of whom *also* 62 deaths occurred – an incidence of 62 per 3162 or 4%. Note: the incidence in the centre group was half that in the top group, but there were the same number of deaths.

So the question remains, do you seek out the group with a 9.1% incidence and treat them, or do you treat everybody? It is a difficult question to answer. It is not a scientific question; it is one of public health, in the same vein as the family who argue that in spite of their obsessive care in dental hygiene in the prevention of dental caries, they are still forced to drink fluoridated water just because others are less obsessive or even negligent in oral hygiene.

Most nutritionists will now accept moderate dietary intervention for the whole population, possibly coupled with some screening programme for those most at risk. Certainly few would go along with the view pertaining in the seventies that the whole nation must switch from a diet dominated by saturated fats to one equal in both saturated and polyunsaturated fats. The leading investigator of the Los Angeles Veterans Administration Trial put it thus:

> When confronted by a patient with genetic hyperlipoproteinaemia* and with a very high serum cholesterol ... we must recognize that risk of coronary heart disease may be very high indeed and that the known and accepted risks of diet modification are acceptable. On the other hand for the general public, or even with patients with a 'high-normal' serum cholesterol ... the urgency is substantially smaller and the argument for conservatism correspondingly greater.

The way forward

So where to from here? How does the world of science respond to the challenge of coronary heart disease. To my mind it goes back to the drawing board or to the laboratory bench and dirties its hands in trying to understand the cellular and molecular basis of the disease. And that is where we are going, back to talk about the low-density lipoprotein (LDL),

* i.e. excess blood cholesterol.

the blood particle bringing cholesterol from the diet and the liver to the tissues.

Why do cells in an arterial lesion break the rules and accumulate toxic levels of cholesterol? Is there a breakdown on the side of the cell? Or is there a breakdown on the side of the LDL? Or both?

One of the most common fallacies, even among so-called eminent physicians (who may know their way round an open-heart operation but not around cells and molecules), is that the more cholesterol there is in blood, the more there will be in the cells exposed to that blood. We have seen that cells regulate LDL uptake and cell cholesterol content so cells do not accumulate cholesterol willy-nilly. At first it was suggested that whilst this regulatory mechanism exists, certain cells (macrophages), which play a key role in heart disease, are scavengers and don't regulate cell cholesterol levels properly. This proposition was tested and the macrophages were found to be quite orderly. They didn't accumulate cholesterol. It was found, however, that you could 'trick' them. If, instead of normal LDL particles, they were offered LDL particles which were cross-linked to form even larger particles, each comprising several individual LDL, they accumulated higher levels of cholesterol. Whilst such trickery is possible in the artificial world of the laboratory it isn't easy to see how you could chemically alter LDL in blood. Now let us turn our attention to the pathology of heart disease.

One key feature in the development of both the arterial lesion and an obstructive clot is that platelets, small saucer-like cells originating in the bone marrow, are involved. Platelets aggregate by way of a stimulus to the platelet which allows the cell to produce a prostaglandin compound which causes aggregation. This substance is called thromboxane A_2. It is incredibly potent at minute doses, as are all prostaglandins (see Chapter 2). For every molecule of prostaglandin synthesized, a tiny chemical called malondialdehyde (MDA) is also produced. This tiny chemical was one of several used to cross-link LDL so that cells were tricked into accumulating cholesterol. So, is it possible that this actually happens at the level of flowing blood in the coronary artery? We should try to find out. If such cross-linking could be proved, we would have a handle on the disease mechanism. If it cannot be proved, we should move on.

Should coronary heart disease ever be licked, I believe it will be because of the pursuit of laboratory, rather than purely epidemiological objectives. Many people talk of prevention of coronary heart disease. Yet that is not what they mean; they mean postponement, so that if heart attacks occur it should be among the elderly, where indeed the majority do occur. What is resented is when young men and women have heart attacks. To die in middle age is seen as a failure, not least by modern medicine. It is a point not properly thought out by those wishing to change our diet to prevent coronary heart disease. Certainly, an 'untimely' death is a great tragedy; but can we all demand three score years and ten as a birthright, and upon what biological reasoning is this demand based? Science may tell us how people die, not why. All the trials and programmes of mass intervention will, I believe, add nothing new to what we know already.

Summary

- Coronary heart disease is caused primarily by a narrowing of the lining of arteries feeding the heart muscle.
- Acute heart attacks occur when platelets form a clot to block completely the flow in an already narrowed artery.
- The lesions which narrow the flow have, among other features, an excessive accumulation of blood cholesterol.
- The higher the level of cholesterol in the blood, the higher the risk of heart disease.
- Dietary changes can lower blood cholesterol and may therefore lower the incidence of heart disease.
- This is most likely if you start off with a very high level of blood cholesterol.
- It also requires drastic dietary action.
- Whilst this drastic dietary action is advisable in the face of elevated risk at high levels of blood cholesterol, it may not alter total death rates among the majority of us in the middle range of blood cholesterol distribution.
- Heart disease is a multifactorial phenomenon involving age, gender, genetics, geography, smoking, blood pressure, exercise and diet.
- The true resolution lies in the hands of the basic medical scientists, hacking away at the problem in their laboratories.

10
Adding and taking away: the modern diet

In the highly mechanised countries, thanks to tinned food, cold storage, synthetic flavouring matters, etc., the palate is a dead organ . . . look at the factory-made, foil wrapped cheese and 'blended' butter in any grocer's . . . Wherever you look, you will see some slick machine-made article triumphing over the old-fashioned article that still tastes of something other than sawdust.

George Orwell: *The Road to Wigan Pier*

The widely held view that our grandparents' diet was better than ours is clearly not new. George Orwell expressed his hankering for bygone fare in 1937. I doubt that he would have enjoyed Edwardian food, unless of course he were to be rich which would have been somewhat out of character! In general, their food was plain monotonous and heavily adulterated. Today's diet, more so than in Orwell's day, is technologically contrived and as such is widely mistrusted. Before exploring this mistrust, it is necessary to look at the technological basis of modern food production.

Our Western society is a technological civilization. Many of our homes are centrally heated or air-conditioned. We can reach any part of the globe with ease by telephone or jet. Our lives are managed by computers, our bodies with drugs. We want and defend this lifestyle and, for as long as we do so, we must accept that food production will be technologically based. It begins on the farm with the use of fertilizers, pesticides, veterinary drugs and hormone implants, and extends to the machinery of milking, harvesting and so forth. From there, the technological input extends to food processing, from traditional drying to modern canning. The technology of storage and distribution follows, and the sequence ends in the home with 'fridges, freezers, blenders, toasters, microwave ovens, hot-pots, percolators, etc. It is possible to be selective in how far a given person wishes to rely on this technology but it is not possible for all of us to shun it all of the time. It is here to stay. Quite simply, if you want your telephone, you have to have your burger. Or, as the Good Book says: 'Give me neither poverty nor riches; Feed me with food convenient for me' (*Proverbs*, **30:8**).

The concern that people have over modern food processing is that the product is either contaminated by undesirable agricultural residues, or is treated with unnecessary additives and is nutritionally inferior to traditional foods. If there are risks attached to modern foods, however, they must be put alongside the benefits. In a similar way we can see the values of electricity, X-rays, penicillin, television and automobile transport, and also their attendant dangers. In evaluating residues and additives, one needs therefore to establish a benefit-to-risk ratio.

Since there is no benefit and considerable risk associated with environmental contaminants of food, there is universal disclaim of their presence. Possible contaminants include heavy metals such as lead or mercury, residues of antibiotics and pesticides, and residues of growth-promoting agents such as cattle hormone implants. The presence of these compounds in food is routinely checked by public analysts, whose capacity to act against malpractice depends on the adequacy of legislation and the severity of the punishment. Most food companies are highly sensitive about the presence of environmental contaminants, for years of delicate fostering of brand image can be shattered by one adverse report. Nonetheless, the public, their representatives and consumer groups need to exercise their vigilance to ensure that residues are minimized and standards are constantly revised.

From residues which nobody intended to be present in foods, we move to food additives which have been added deliberately. The risks and benefits vary considerably, and the additives can best be classified on the basis of their benefit: obviously beneficial, of doubtful benefit but of some convenience, and of no obvious benefit.

Almost all food is perishable, and central to the survival of man has been the capacity to preserve food obtained in times of plenty for use in leaner seasons. Drying has been a widely used traditional method for preserving fruits, cereals and meats. Salting has been used for centuries to preserve

The preservation of food has always been an important concern; in the past, preserved food provided about the only way of surviving a long period of famine.

Francis Bacon (1561–1626) – the distinguished philosopher and statesman – conducted an early experiment with food preservation techniques, though with tragic consequences: he died of a chill soon after stuffing a chicken carcass with ice and snow in the depths of winter in an attempt to preserve it from decomposition.

food (see Chapter 5). Pickling is another traditional process, as is fermentation. Modern methods include canning, freezing and accelerated freeze-drying. However, it is still necessary to use additives to preserve certain foods for otherwise their production, distribution, storage and retailing would require rigorous refrigeration procedures which could not realistically be relied upon to ensure safety.

The main food preservatives in use are: the nitrates and nitrites, used in such meat products as bacon and pressed meats; propionic acid, used in baking and confectionery such as cakes and bread; sulphur compounds such as sodium metabisulphate or sodium sulphite, used in wines, beers, sausages, fruit juice, pastes, purées and sauces; and benzoic acid, often used as a substitute for sulphur-based preservatives. Without these preservatives, the foods listed would show rapid spoilage due to the growth of bacteria, yeasts and moulds.

It is worth considering the incidence of food-borne diseases in the USA, a country heavily reliant on processed food, additives included. In 1978, 5000 cases of established food-borne disease were recorded. Only 1% were due to chemical agents, i.e. pesticides, heavy metals, etc. Of the remaining bacteriologically related food-borne diseases, 1% were due to the highly dangerous bacterium *Clostridium botulinum* (i.e. botulism), 40% were due to *Salmonella* organisms, 27% to *Staphylococcus aureus*, and the rest to a variety of microbes and parasites. These represent only the reported and verified cases. A further 6000 cases were recognized as being food-borne but the causative organism could not be identified. Estimates of the total number of cases of food poisoning, including those not reported on the basis of their being mild or short-lived, are in the region of 1.4–3.4 million per annum for the USA, about 1–3% of the population. Consider how much higher these numbers would be without preservatives being added to foods. So, given our present system of food production, the use of preservatives offers considerable benefit to the consumer. What of the risks?

Of all the preservatives, the nitrates and nitrites have attracted most attention. They are used in those situations which favour the growth of the botulism organism and are thus of considerable benefit; they also have an attendant risk. Nitrites are the active factor in inhibiting the growth of *Clostridium botulinum*. They are provided either directly, or indirectly as nitrates which are converted to the active nitrite form by bacterial action in the food or in the gut. Nitrites, reacting with amino acids derived from dietary protein in the acid environment of the stomach, form nitrosamines and some nitrosamines have been shown to be highly carcinogenic in test animals. However, toxicologists quickly point out that the nitrosamines used in the cancer tests are present in vastly excessive amounts, well above what people might be exposed to. Nonetheless, the concern is there, and is bolstered by two epidemiological observations. Japan and Chile have the highest rates of stomach cancer in the world: the Japanese have a very high intake of dried, salted foods which are rich in nitrite; and the world's largest nitrate mines are in Chile.

Where does this leave us? For one thing, we must consider the contribution of nitrates from non-food sources to overall intake. The average consumption of nitrites in the USA is 11 milligrams per day and that of nitrates is 100 milligrams per day. Cured meats provide the major source of dietary nitrites, contributing 21.2%. Virtually all of the remainder comes from saliva! This salivary nitrite arises from salivary nitrate which is, in turn, derived from dietary nitrates. Cured meats contribute only 9% of nitrate in the diet; vegetables account for 86%. Vegetables absorb nitrates from soil moisture for conversion into plant protein and, in so doing, accumulate a nitrate reserve. Hence, even if cured meats were excluded from the average American diet, the mean daily intake from the food chain would be 90 milligrams of nitrate, 2.2 milligrams of nitrite, with a further 8.6 milligrams of salivary nitrite derived from dietary nitrate! Clearly, if nitrosamine formation in the stomach increases the risk of cancer, removing nitrates and nitrites from processed food would have little effect. Given that these compounds inhibit the growth of such a deadly bacterium as *Clostridium botulinum*, the case for leaving them alone seems clear.

A second group of additives in this category are the antioxidants; these chemicals are added to fats and oils to prevent them from going rancid which, aside from causing ghastly smells and tastes, would lead to nutritionally undesirable and potentially toxic compounds. The reaction involves atmospheric oxygen and fatty acids and its prevention is mediated by antioxidants. These antioxidants occur in nature, for this is the main function of vitamin E, to ensure that the highly polyunsaturated fatty acids in membranes do not react adversely with oxygen in the cell. So the term ‘permitted antioxidants’ appears on many food labels. So far nobody has mooted these compounds as being in any way dangerous.

The category ‘of doubtful benefit but of some convenience’ includes emulsifiers, stabilizers, anticaking agents, acids, non-stick agents, air-excluders, phosphates, humectants, sequestrants, firming-agents and antifoaming agents. We can consider a few of these. Emulsifiers help to keep a water–oil homogenate stable, thus ensuring that mayonnaise does not separate into its two liquid components. Of course, you could shake the bottle every time you use it, but that would be inconvenient. Similarly, you could put your salt into an open salt cellar, complete with spoon, and break up any lumps with your fingers. That too is deemed inconvenient, so non-stick agents are used. Cake icings ought not to dry out, so humectants are used. Many of the additives in this group are naturally occurring compounds. Citric acid, as in lemons, is a popular acid. Magnesium carbonate, used as a non-stick agent, occurs naturally, as does the humectant sorbitol, which is digested and metabolized. Glycerol monostearate is used as an emulsifier – and is produced routinely from natural fats during digestion; because the effective benefit of these compounds is marginal, however, the compound could be withdrawn if there were any hint of a deleterious effect. It should be re-emphasized that the principle of using chemicals for these functions in foods is fundamentally acceptable in our technological diet.

Flavouring and colouring agents are difficult to classify into any of the three nominated categories, so let us assume them guilty until proven innocent! All flavouring and colouring agents are of no obvious benefit to the consumer. When peas are canned, they lose their green colouring and become greenish-grey/greenish-yellow; strawberries become straw-coloured. The manufacturers of canned products allege that if they did not restore the natural colour with non-natural chemicals, we would not buy the products. They are probably right and they cite one study where sales fell by 50% when one manufacturer decided to try leaving them out. We 'want' tinned peas to be green, but the truth is that we have been told that is how we want them by the advertising agencies. 'Fresh', 'green', 'farm-fresh' and 'golden' are the language of the advertiser. So, even though there is nothing 'fresh' or 'green' about tinned peas, we are conditioned to think that there is. You may decide yourself whether the benefits of food-colouring agents are worth the risks.

The same might be argued for flavouring agents. Potato crisps come in a whole range, from plain (ready salted), to those flavoured with an array of wholly synthetic agents, ranging from prawn-flavoured to barbecued chicken-flavoured crisps. Flavouring agents do not have to be synthetic; salt and sugar are probably the most widely used. For example, salt is added to tinned green peas, and sugar to some yoghurt. Fresh green beans contain only minute quantities of sodium (1–5 milligrams of sodium per 100 grams of fresh beans) tinned green beans contain 100 milligrams of sodium per 100 grams of finished product. Fresh cow's milk contains no sucrose whatsoever. Most yoghurts, a food with a 'natural' image, contain sugar.

To my mind, none of this latter category of additives is beneficial in the physiological or hygienic sense, but they may have some gastronomic benefit. This benefit is therefore marginal, such that any associated risks tip the risk-to-benefit ratio into unfavourable quantities. In general, the risks associated with colouring agents or synthetic flavourings can be divided into risks of cancer and risks of hypersensitivity. To prove beyond all reasonable doubt that an additive is carcinogenic, i.e., cancer-causing, is nigh impossible. How much should be fed to, or injected into, or sprayed onto the rat or rabbit or mouse? A convenient figure of one part per million or multiples of ten thereof? Or perhaps the amount normally added in the appropriate food? Suppose an additive 'A' were normally added to jams at the rate of 10 parts per million and, at that level, was suspected of causing tumours but at lower levels had absolutely no effect? Should 'A' be banned? Is there anyone who eats only jams? Surely jams are spread on bread or confectionery at some meals only and only on certain occasions. Under these circumstances, the concentration of 'A' in jams bears no relation to the normal intake of 'A' and, therefore, to the level of 'A' in the diet. In the end, the decision to permit or prohibit the use of an additive is made by the consensus of selected experts who have sifted through mountains of wholly inadequate and inconclusive data; such experts usually operate by the simple rule, 'if in doubt, throw it out'.

Hypersensitivity is a much more difficult issue, largely because the complaint is located in the suffering individual. The great majority of people do not get bizarre behavioural symptoms on eating processed food. The fact that some do presents a problem, not of whether to ban or not to ban, but of how to label foods. The use of vague terms such as 'permitted antioxidants' suits the manufacturer who does not want to use such chemical terms as butylated hydroxytoluene, but to the hypersensitive individual, detailed knowledge gleaned from food labels could prevent some nasty symptoms. Now, codes are being used, for example, stabilizers E410; preservatives E211; colours E123 and 124.

Some compounds are deliberately added to foods to bolster sales, to cash in on, or even help to create, the nutritional awareness of consumers. These are the vitamins and minerals. Such and such a breakfast cereal can provide

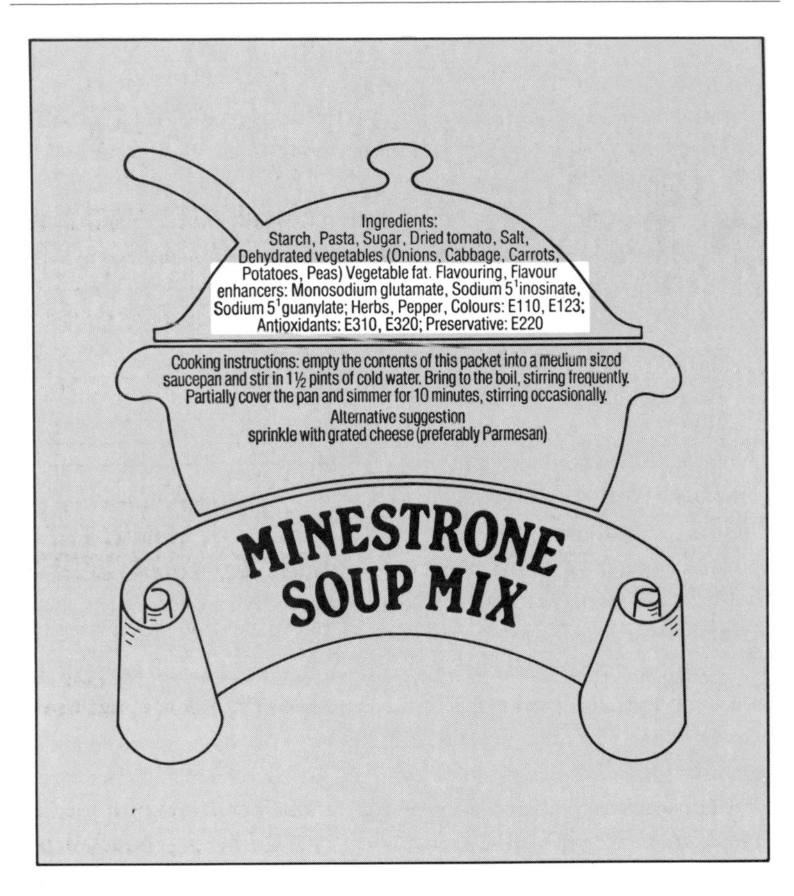

There is nothing mysterious about E additives. They are simply chemicals added to help colour, flavour or preserve foods (in this example, packet soup). They have all been tested for safety and instead of listing their full chemical names on the back of food packaging it has proved convenient to use an internationally recognized shorthand version of their names.

your child with half its daily requirements of this vitamin or that mineral, implying that if not served religiously at breakfast our children would suffer nutritional deficiencies. These are simple marketing techniques, just as is describing a soft margarine as being made from pure natural oils when, in fact, it may contain hydrogenated fish oil, refined and deodorized. Then there are fruit juices made from 100% fruit. Well yes, but what is the ratio

(*a*)

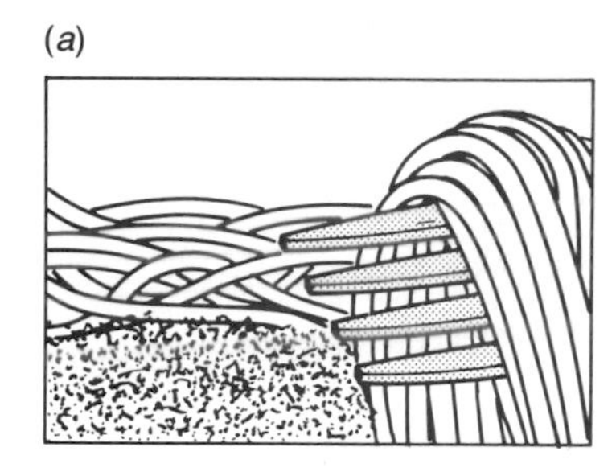

NUTRITIONAL INFORMATION PER 100G UNCOOKED* PASTA	
ENERGY	– 322K cal/1352Kj
DIETARY FIBRE	– 10g
PROTEIN	– 11g
FAT	– 1g
CARBOHYDRATE	– 72g
IRON	– 1.3mg
VITAMINS	
THIAMIN	– 0.14mg
NIACIN	– 2.0mg

*Equivalent to 300g of cooked pasta

(*b*)

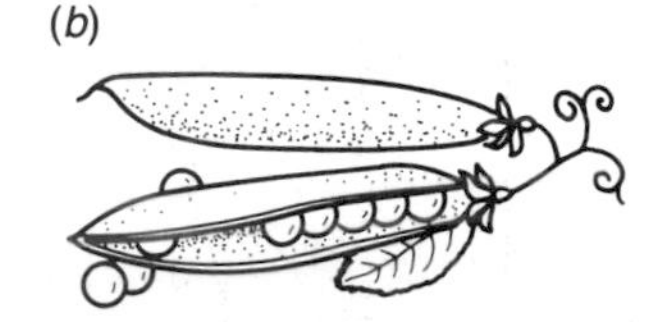

NUTRITION

Garden peas are a good source of dietary fibre. A serving also contains one-third of the daily requirement of Vitamin C which we need to ensure general good health.

	BOILED	
AVERAGE COMPOSITION	PER 75g (2¾oz) serving	PER 100g (3½oz)
Energy	130 kJ 31 kcal	172 kJ 41 kcal
Fat	0.4g	0.5g
Protein	4g	5.5g
Carbohydrate	3g	4g
Fibre	9g	12g
MINERALS/ VITAMINS	% RECOMMENDED DAILY AMOUNT	
Vitamin C	33	13mg

THIS PACK CONTAINS APPROX 6 SERVINGS

INFORMATION

The modern trend towards more informative labelling of a food's composition is to be welcomed. It enables the consumer to see at a glance how much fat, fibre, protein, etc is present and provides the basis for making an informed decision towards an improved diet. These two labels come from (*a*) high-fibre pasta and (*b*) peas.

of peel, to pulp, to juice? Advertising food is just as sophisticated as advertising deodorants or hi-fi equipment, or any other product of our industrial technostructure. It must be said that while most mineral and vitamin supplements are added to help sell a food, some are added to meet legal requirements, as are vitamins to margarine or minerals to white flour, depending on where you live.

If all of this has left the 'organic' food enthusiast smiling smugly, be patient and read on. For if there is one yardstick against which the risks of food additives can be evaluated, it is the group of compounds which occur naturally in foods, which have no nutritive value (being intended for the needs of the plant and not the consumer) and which carry a risk to the consumer. These are the naturally occurring food toxins. The fact that we cope with these is irrelevant. What matters is that if they did not occur naturally, and a manufacturer for whatever reason, real or otherwise,

The Brassica vegetables provide an educative example to those who think that food additives are an unnecessary evil and that pure, unadulterated food is best. The brassicas, such as cabbages and sprouts, contain a mildly toxic anti-nutritional factor that can produce the deficiency disease goitre. Fortunately, for those on an adequate diet this effect is only of academic interest, but for people living on inadequate diets the effect can be real. Other natural toxins are discussed in the text.

wanted to add them to a food, he simply would not be allowed to. The fact that these toxins and other compounds in foods are destroyed by soaking and cooking would not mitigate in their favour.

The first group of naturally occurring food toxins is the antinutritional factors. About 200 million people suffer from goitre, the disease of iodine deficiency which is characterized by a swelling of the neck as the thyroid gland expands in its attempt to make the iodine-rich hormone thyroxine. Most affected individuals live in regions where iodine is poorly available in the soil, but some 8 million of them develop goitre because their foods contain goitrogens, compounds which impede thyroxine production. The

Brassica vegetables are the worst offenders (i.e. cabbages, kale, Brussels sprouts, etc.). A less-serious antinutritional factor is phytic acid, widely distributed in cereals and legumes. Phytic acid forms complexes with calcium, iron, zinc and other minerals and, in so doing, reduces their bioavailability. Indeed, this has been the basis of the fortification of flour with calcium and iron, which was introduced during war-time rationing in many countries.

The next group of nasty natural chemicals are the toxins themselves. Most beans, especially soya beans, contain an anti-trypsin factor. Trypsin is secreted by the pancreas into the gut to digest dietary protein. The anti-trypsin factor, or trypsin inhibitor, binds the enzyme and, as a result, the pancreas compensates by an even further trypsin secretion. Finally, the pancreas ceases to function and pancreatitis is deemed to have occurred. Heating destroys the anti-trypsin factor as it destroys most enzymes. It is still of some concern, however, for many infant formulas are based on soya bean flours and, unless rigorous processing and checking occurs, infantile pancreatitis can occur. Other toxins in legumes include lectins, which cause gastroenteritis when raw or improperly cooked red kidney beans are eaten, and haemagglutinins, which cause red blood cells to clump. Lathyrism is a crippling disease of the nervous system which is still quite widespread and is brought on by eating certain varieties of vetch. Up to 7% of the population of certain regions of India suffer from lathyrism. Cyanide poisoning can occur after eating about 125 grams of poorly prepared lima beans because these beans contain about 100 milligrams of releasable cyanide per 100 grams of beans. Cassava, widely consumed in the Third World, also contains releasable cyanide and studies in West Africa, Jamaica and Malaysia have suggested that tropical diseases of the eye (amblyopia) and nervous system (ataxia neuropathy) may arise from continuous ingestion of such foods. The list is legion but might perhaps be concluded with the ludicrous: nutmegs, parsley and carrots contain myristicin which can induce hallucinations! Adding myristicin to potato crisps would never be allowed!

Perhaps the nastiest naturally occuring toxins are the mycotoxins, produced by certain moulds which infect seeds, not through agricultural incompetence or ignorance but as a matter of course and often in spite of agricultural efforts to avoid infection. ‘St Anthony’s Fire’, a serious

The controversy about food additives can perhaps be seen in a clearer light by considering such basic foods as nutmegs, parsley and carrots – all of which contain small quantities of a hallucinogenic chemical that would most certainly not be allowed to be used as a food additive.

problem in the Middle Ages, is associated with an ergot-producing mould growing on cereals. Today the biggest worry are the aflatoxins, produced by the mould *Aspergillus flavus*; it is to farm livestock that the greatest threat occurs. In the UK in the early 1960s, 100 000 turkeys died when peanut meal protein was found to be contaminated with aflatoxins. Fortunately, man seems to be able to cope with these moulds or at least with *A. flavus*. A dose of 300 milligrams aflatoxin per kilogram body weight given to rats caused severe liver cancer in 100%. Thai villagers, habitually consuming 75 milligrams aflatoxin per kilogram body weight, had a low incidence of liver cancer – 12 per 100 000.

So much for so-called safe organic food. The facts are that every non-nutrient in food, be it natural or synthetic, present normally, deliberately added or inadvertently contaminating, constitutes a risk. In most instances the risks are small and leave no reason why we should not enjoy all foods and, for those of us lucky enough, be grateful for that enjoyment; there are many who would swap places with us.

The very last point to be considered is the view that modern food processing has produced a product nutritionally inferior to the natural, unrefined or unprocessed food. This is partly true. It is particularly true of dietary fibre: during the period following the Second World War, fibre became unfashionable, refined foods were sought after and fibre intakes fell. Fortunately, most of the food producers have responded to the revived interest in fibre as a desirable nutrient by providing a wider range of produce to include wholemeal breads, pastas, etc.

Little if any adverse effects on the nutritional values of food proteins, carbohydrates or fats occur during modern production and processing. Some vitamin losses do occur with heating during canning, but the same losses will occur in the kitchen during cooking. All in all, tinned or frozen produce is as nutritionally adequate as fresh produce. The notion that chemical fertilizers differ from organic fertilizers is naïve. Plants absorb nitrates, sulphates, phosphates and other such ionized elements and it does not matter to the plant whether the sulphate was produced by an industrial chemical process or by rotting compost. A sulphate ion is a sulphate ion.

11
Fads and fallacies

Here is an extract from the list of diseases which are covered by Adelle Davis in her multimillion-selling book *Let's Get Well*. In each instance, nutritional intervention is prescribed to prevent or cure:

> Abdomen, water blisters on; abdomen, cramp; abscesses; Achilles tendon, shortening of; acid, stomach; acid urine; acidosis; acne; Addison's disease; adenoids, enlargement of; ageing; air sickness; anaemia; allergies; anaphylaxis; aneurysms; angina; anus, itching of; appendicitis; arthritis; asthma; atherosclerosis; athlete's foot, . . .

It isn't that Adelle Davis never got past 'A' in her medical dictionary, just that I rest my case there. Certain people regard poor nutrition as the cause of almost every ailment and are just as likely to turn to diet for a cure.

The problem which professional scientists face in evaluating such unsubstantiated claims – which range from the plausible to the incredible – is that many of them have the ring of truth. It isn't, to the average man, utter tripe at first glance. If it were it wouldn't succeed. Some of the more outlandish claims are, however, difficult to swallow, no matter how convincing the sales pitch, e.g. 'autoimmune urine therapy' (see page 94).

Half-truths need to be ruthlessly exposed and yet the campaign should not become an inquisition in which valuable conjecture is lost or its investigation delayed. Science and medicine operate on a system of dogmas or, more correctly, 'widely held hypotheses'. None of these hypotheses is sacred, otherwise Newton would not have replaced Copernicus, nor, in turn, have been replaced by Einstein as the high-priest of prevailing physics. However laboured the change in a 'widely held hypothesis' may be, it must progress along the lines of scientific pursuit, of bold conjecture and austere refutation. Anecdotal data may be useful in formulating a good hypothesis, but a good hypothesis is only as good as its last test. Too many self-professed experts on nutrition don't adhere to this rule. They don't test to satisfactory standards.

The range of misinformation and half-truths in nutrition is daunting, covering everything from diets and diet supplements to make one lose weight rapidly, to toxic compounds labelled as vitamins and given to vulnerable cancer patients. We shall consider just a few examples.

Obesity

Cellulite is a tissue of dimpled fat around the thighs, a form of fat that is resistant to dietary manipulation. Well, that's what the people who sell the means of shedding cellulite say. But no text-book on histology lists cellulite in its index. The term is a convenient invention and a useful gimmick for advertising and publicity. It is not indeed in standard texts on histology, nor in the extensive current research listing *Index Medicus*. Cellulite represents just one extreme example of the sort of misinformation that obscures the painstaking scientific study of obesity. In making this unfounded anatomical claim, it overrides in one fictitious word the problem people have in assuming the advertiser's image of thighs. It's not fat, it's cellulite. If all fat *can* be lost by dieting, what sort of diet should be followed? Just about every conceivable angle has been tackled.

The low-carbohydrate or carbohydrate-free diets were very popular in their day. Rapid weight losses do occur and have been documented. When carbohydrate in the diet is restricted or absent, blood glucose has to be maintained from other sources, notably the amino acids of protein. Equally, nearly all the energy derived after the initial few days will be from fat. This causes a metabolic conflict. An intermediary in the process of fat

A montage of a few of the claims that some adverts would have us believe.

oxidation is also an intermediary in glucose synthesis from protein. Alas the paths go in opposite directions, so a compromise is reached. Glucose synthesis continues and fat oxidation continues but is only allowed to proceed to partial completion. The partially combused fats form ketone bodies which the proponents of the carbohydrate-free diet say is the basis of its success, as the ketone bodies are excreted in the urine. So you only need to half-finish the burning of fat in the body; the rest goes down the loo. To make the job a success, it is suggested that dieters buy strips of indicator paper from their chemist to detect urinary ketone bodies and adjust their diet to maintain an adequate rate of excretion. Ketone bodies *are* excreted in excess on such diets but would contribute little to fat loss. A reasonable flow of ketone bodies in the urine would mean a loss of only 50 kilo calories per day to the dieter. To lose 500 grams of fat at this rate would take ten days. So why the weight loss? There are two reasons, which are interlinked.

The muscles of the body, and to a smaller extent the liver, store glucose as glycogen. A carbohydrate-free diet rapidly leads to the breakdown of this glycogen to meet the ever-present need for blood glucose. Blood glucose will also be derived from amino acids. Each molecule of glycogen is linked to four times its weight of water. Loss of glycogen will lead to loss of body water and of body-weight. The second contributory factor to weight loss also involves loss of water, for a low-carbohydrate diet leads to the loss of sodium in the first week or so. Depletion of body sodium necessarily requires loss of water to maintain osmotic balance, so the carbohydrate-free diet leads to a rapid loss of water brought on by glycogen depletion and sodium excretion.

These diets are not without danger. In the Korean War, a group of Canadian soldiers living almost exclusively on corned beef for a period developed symptoms such as listlessness, fatigue and light-headedness. Diabetics have similar symptoms. These are classical symptoms of hypoglycaemia (low blood sugar) and ketosis (excess blood ketone levels), both of which arise on pursuing a carbohydrate-free diet. Other symptoms include nausea and halitosis. Alcohol exacerbates matters. All in all, the pursuit of a carbohydrate-free diet is not wise. Neither is its corollary, carbohydrate excess. This advocates a low-protein diet, with an intake of less than 20 grams per day compared with a Western average of 90 grams per day. This diet aims to trim bulky muscles in the pelvic, thigh and shoulder areas. Unfortunately, when dietary protein is restricted, all protein-rich tissues are affected – muscle, heart, liver, spleen, etc. Digestive capacity is reduced and the immune system is impaired. Again, not a wise diet.

A variety of high-fibre diets have been advocated, including the Pritikin Programme, emanating from California, and the 'F-plan diet'. Foods such as meat, milk and eggs are virtually totally digestible; the dry matter is comprised almost exclusively of protein, fat and carbohydrate, all of which yield energy. Vegetables, fruit, cereals and legumes contain fibre (see Chapter 3) which, being indigestible, 'dilutes' these calorie-yielding components. These diets tend to be relatively low in protein and this has been a criticism of them. This is probably unfair since many millions of people in many parts of the world live perfectly healthy lives on such diets, some by choice, some by necessity. Indeed a shift in dietary habits towards a more vegetarian-type diet is, in practice, the likely consequence of the views of prevailing gurus in nutrition, a matter to be dealt with in the final chapter.

Whilst some 'wonder diets' are concerned with varying the relative proportions of protein, carbohydrates and fat in the diet, others turn their attention to specific foods and adjuncts of slimming. The grapefruit has somehow earned a reputation among slimmers. It is alleged that grapefruits contain 'enzymes' which somehow bring about the mobilization of body fat. This unique citrus fruit plays a central role in some diets. Just consider what the grapefruit is supposed to do: 'dissolve' body fat by way of a unique fat-dissolving enzyme it contains. All enzymes are protein. All proteins are digested in the stomach and small intestine and are absorbed as amino acids. If tiny amounts did cross the gut intact, the immune system would be called upon to hustle out these intrusive foreigners (see Chapter 8). Some proteins, such as antibodies in breast milk, do resist digestion but they are not absorbed: they're excreted intact in the faeces!

From the fiction of the grapefruit's fat-dissolving enzymes, we move to yet another form of dietary delusion, also requiring the action of a specific protein: starch blockers, proteins derived from certain legumes which inhibit amylase, the carbohydrate-digesting enzyme. Starch blockers permit the most ardent slimmer to wolf into pastas without worry. Right? Wrong! Starch blockers are proteins, and proteins, as we saw above, are digested, beginning in the stomach. Carbohydrate digestion, though, doesn't seriously begin until the small intestine. So it is likely that by the time amylase swings into action the starch blocker is half-way digested. And even if the starch blocker worked, all that starch would wind its way down to the bacteria in the lower gut, whose ensuing field day would pin the would-be-slimmer to the loo with diarrhoea and flatulence.

One of the most bizarre slimming adjuncts is the amino acid ornithine. It is alleged to alter body composition by stimulating human growth hormone (HGH) and leading simultaneously to the build-up of muscle and breakdown of fat. It is a classic example of yet another half-truth concerning dieting. It is true, that ornithine, and its 'colleague' amino acid argenine, stimulates HGH when given intravenously. The *minimum* effective dose of argenine is 30 grams intravenously over 30 minutes! To achieve that influx of argenine into the blood by dietary means would require a massive ingestion of argenine which would have most undesirable effects in creating amino acid imbalance. Ornithine is more effective than argenine in stimulating HGH but, by my calculations, the minimum effective oral dose would be at least 30 grams, equivalent to 120 tablets each containing 250 milligrams of ornithine. And there's a further catch which the proponents of ornithine ignore. The stimulatory effect of these amino acids on the release of HGH is dependent on the female sex hormone oestrogen. Their effect is confined to women but that hasn't prevented the displaying of Mr Universe persons next to ornithine tablets in some advertising literature! Are these he-men taking sex hormone pills? Would excess HGH automatically lead to a Mr Universe figure, even ignoring the need for hormone replacement therapy to create such an excess of HGH? There is a 'natural' experiment. Certain people are born with an excess of HGH; they have a characteristic facial appearance with a protruding chin and forehead. HGH does lead to muscle growth in such individuals but also to enlarged hearts, livers, spleens, lungs, guts, kidneys, etc. In other words, HGH stimulates protein synthesis and does so in all protein-rich organs.

The bottom line in slimming, and it has been dealt with elsewhere (Chapter 4), is this: if energy intake exceeds energy expenditure, you gain weight; not necessarily in proportion to the excess, but you do gain weight. Conversely, if energy intake falls short of energy expenditure, you lose weight, more or less in proportion to the deficit. Energy expenditure has many forms, the two most important being resting metabolism (the energy the body needs just to stay alive) and exercise. Little can be done to raise the former. Exercise, combined with a sensible calorie-controlled diet, is the answer.

Vitamins

In 1969 the Food and Drugs Authority, the foremost US body on such matters, carried out a survey of health practices and opinions in the USA and found that most Americans believed that: extra vitamins will provide more 'pep' and energy, people who feel tired and run down probably need more vitamins, and most Americans do not eat balanced diets. There are two conclusions one can draw from this survey: that Americans and presumably many other nationalities are ignorant in matters of diet and that this ignorance is readily taken advantage of by the promoters of these products. In their excellent book *Vitamins and Health Foods: The Great American Hustle*, Victor Herbert and Stephen Barrell point out just how these misunderstandings are played upon. They list seventeen claims that are a tell-tale sign that your gullibility is being tested and probably exploited:

1 He advises you to buy something which you would not otherwise have bought.
2 He says that most disease is due to a faulty diet.
3 He suggests that most people are poorly nourished.
4 He tells you that soil depletion and the use of 'chemical' fertilizers results in less nourishing food.
5 He claims that modern processing methods and storage remove all nutritive value from our food.
6 He tells you that under stress, and in certain diseases, your need for nutrients is increased.
7 He says you are in danger of being poisoned by food additives and preservatives.
8 He says that if you eat badly, you'll be okay if you take a vitamin or vitamin mineral supplement.
9 He recommends that everybody take vitamins or health foods or both.

10 He recommends a wide variety of substances similar to those found in your body.
11 He claims that natural vitamins are better than 'synthetic' ones.
12 He warns that sugar is a deadly poison.
13 He tells you it is easy to lose weight.
14 He promises quick, dramatic, miraculous cures.
15 He uses anecdotes and testimonials to support his claim.
16 He'll offer you a vitamin that isn't.
17 He espouses the 'conspiracy theory' and its twin the 'controversy claim'.

Many of these items will be dealt with in this chapter, while others are discussed elsewhere.

Vitamins do not function in a magical manner. They act biologically, each having its own specific function. For a particular biological reaction to proceed at the correct pace, there is a given quantity of the vitamin required. Excess vitamins will not push the pace faster, just as having a full tank will not make a car go faster than its engine capacity will allow. Fuel deficiency will of course be problematic. Similarly, inadequate intake of vitamins will slow down the reactions for which they are needed. Therefore, one needs to know the levels of vitamins required to satisfy the body's requirements. Thereafter one is concerned only in preventing deficiencies and exposing the mistaken, and at times dangerous, practice of megavitamin therapy. Much of this has been dealt with in Chapter 7, and I shall not dwell here on detail except to repeat that most people eating a balanced diet *will not* encounter vitamin deficiency and that vitamin supplements *do not* provide 'pep' or energy.

What I do want to discuss here is point 16 of the checklist profile; vitamins that aren't. In 1949, Ernest Krebs Senior and Ernest Krebs Junior were jointly awarded a US patent for a *method* of preparing a compound from certain seeds which would be useful for the relief from and prevention of 'asthma and allied diseases . . . eczema . . . arthritis, neuritis . . . affections of the skin, respiratory tract, painful nerve and joint affections and even cell proliferation'. Later they declared this substance to be pangamic acid (*pan* = universal; gamic = of seed) or 'vitamin B_{15}'. The basis of their claim that vitamin B_{15} was in fact a vitamin was that it is found invariably where all other B vitamins exist in the plant world. So is water, but not even the Krebs would have the cheek to call water a vitamin. The

Food and Drugs Authority have declared pangamic acid, vitamin B_{15}, 'illegal'. At least three Federal courts have prohibited its sale, judgements which were surely influenced by the discovery of the cancer-causing properties of some ingredients sold as vitamin B_{15}.

The same Krebs family duo marketed another vitamin which wasn't, and still isn't: 'vitamin B_{17}'. Also sold under the trade-name Laetrile, it is the compound amygdalin, found in the kernels of several fruits and seeds. It is 6% by weight cyanide and is highly dangerous. One 11-year-old New York girl died after taking five tablets (250 milligrams Laetrile each), and there have been several other fatalities. Krebs recommended vitamin B_{17} for the treatment of cancer. He was convicted in 1980 by a San Francisco court and sentenced to 6 months in jail. Dr Krebs received his Doctorate, presumably in science, from the American Christian College, which incidentally does not have a science department!

However, the prize for vitamin salesmanship must go to Silent George of Shawneetown: Some years ago, an imaginative promoter in the Southern part of Illinois, USA, calling himself Silent George of Shawneetown, had a good thing going. Removing the labels from small cans of condensed milk, Silent George sprayed the cans with gold paint, then affixed new labels bearing the name Swamp Rabbit Milk. The labels described the cans' contents as 'a balanced product for unbalanced people, rich in vitamins J, U, M and P', indeed much richer in the last of these vitamins than beer or watermelon. Like many labels, these bore a warning aimed at women, slyly suggesting the product's indications for use: 'Do *not* imbibe the potent fluid in the absence of your husband, sparring partner, boy friend or running mate, as the action is fast and it is two jumps from a cabbage or lettuce picnic to a cruise down the Nile with your dream version of Mark Anthony'.

Minerals and trace elements

Food faddists often argue that our diets are inadequate in trace elements because intensive agriculture functions on depleted soils. This is untrue. Like vitamin deficiencies, mineral deficiencies occur in well-defined groups (see Chapter 6) in society and, as with vitamins, excess minerals can be toxic. Minerals such as zinc and selenium are sold in vast quantities to the ill-informed public, usually to offset ageing or promote male fertility. The promotion of zinc as a dietary supplement does provide an amusing and educational example. Superoxide dismutase (SOD) is an enzyme found in many types of cells which for one reason or another generate peroxides. Since peroxides are toxic, they must be rapidly metabolized. That is the job of SOD. It is a copper- and zinc-dependent enzyme, so that literature on zinc often cites this vital function. Would supplementary zinc increase SOD activity above normal levels? No, not at all. So, the promoters can try another approach by selling actual SOD tablets. Now SOD is an enzyme. All enzymes are proteins. All proteins are digested (heard this before?!) and are absorbed as constituent amino acids for the body to make into whatever proteins it decides are necessary; never what the menu decides. There are other instances of this lunacy not related to minerals. For example RNA, ribonucleic acid, is the genetic template for protein synthesis. In the belief that RNA supplements will lead to extra protein, RNA tablets are purchased by gullible people who ingest, digest and then excrete them without any benefit and with some risk. Excess nucleic acids in the blood can lead to uric acid crystallizing out in joints causing gout (see page 144).

Other aspects of nutritional fallacies

Most health food shops stock an array of amino acids. Quite why anyone would buy them is baffling. The average Westerner probably eats from 50 to 100% more protein than he or she actually needs to replace the protein lost (as urea) in urine each day, the excess being used for glucose synthesis and for energy. There are absolutely no instances of normal Western dietary practices which lead to amino acid deficiency. In fact, taking an individual amino acid supplement can be toxic. In protein synthesis, the cell needs all the amino acids to be present in the required ratio. If this condition isn't met, protein synthesis is impaired. The most pronounced effect of amino acid supplements, though, is depression of appetite. People who buy amino acids run this risk, but I have yet to see an accompanying warning on the label.

There is an additional point to be raised here. It is often claimed that certain pharmacological attributes can be ascribed to certain nutrients.

Tryptophan, another amino acid, is sold to induce sleep and reduce depression. If, and it is a big if, it does have these effects, then the compound is not being used as a nutrient, to meet a nutritional need, it is being used as a drug, to meet a pharmacological need.

The conspiracy theory

Many self acclaimed 'experts' on nutrition and health, including Adelle Davis, believe that scientists and doctors stubbornly adhere to conventional ideas for a variety of reasons, usually including their fear of being superseded or of the mystery being taken out of medicine, or the belief that they are being paid to hold these views. Ms Davis writes: 'Though most scientists are dedicated, scrupulously honest individuals, a few doctors, well paid by the vested interests, have become vocal in proclaiming the excellence of the American diet'. This is unfair. Those who proclaim the 'excellence' of the American diet are simply pointing out that the 'deficiencies' on which quacks thrive do not occur in the normal American diet. These same experts do accept that while deficiencies may not occur, excesses in calories, sucrose, alcohol, cholesterol and fat certainly do.

12
A miscellany of matters

Nutrition in pregnancy

Pregnancy can be divided into two phases. The first phase concerns itself with the build-up of maternal nutrient stores and maternal adaptation to conception; the second phase is dominated by foetal requirements. At 20 weeks, midway in gestation, maternal stores of nutrients have peaked. Yet at this point, the foetus is only 10% of its eventual birth weight. During the final 20 weeks, the foetus grows at the expense of maternal nutrient stores. At the same time the mother is depositing stores of fat which will be required subsequently during lactation. Each of these features, the growth of uterine tissues, foetal growth and fat deposition, contributes to the weight gain in pregnancy. A satisfactory pre-pregnancy weight and satisfactory weight gain during pregnancy leads to adequate birth weight and low incidence of 'low-birth-weight babies'. Conversely, an inadequate pre-pregnancy weight and inadequate weight gains during pregnancy lead to smaller babies at birth and a higher percentage of 'low-birth weight babies'. Does this mean that conscientious mothers-to-be should fatten up prior to conception and 'eat for two' during pregnancy? Not really. From a wide variety of studies of nutritional supplements in pregnancy it would seem that in the low socioeconomic groups in underdeveloped countries, nutritional supplementation in pregnancy is valuable; in the industrialized countries, only a small minority could so benefit. Nonetheless, inadequate diet in pregnant women in industrialized countries is undesirable as has been pointed out with regard to folic acid (Chapter 7) and iron (Chapter 6).

Breast feeding

A pair of substantial mammary glands have the advantage over the two hemispheres of the most learned professor's brain in the art of compounding a nutritive fluid for infants. Oliver Wendall Holmes

There is just one error in the above pronouncement. Satisfactory lactation is definitely not related to the size of the lactating mammary gland!

Two hormones play a crucial role in lactation. Prolactin is involved in the synthesis of the components of milk, and oxytocin is involved in the let-down or milking reflex. This latter aspect is also related to such factors as maternal stress and fatigue and to the infant's sucking ability. For all of these reasons, not every mother who attempts breast feeding will be successful, although all who can breast feed should do so. Breast is best!

Milk secreted in the first 72 hours or so after birth is referred to as colostrum. It is richer in protein, particularly proteins which are antibodies. The young infant is born with an adequate supply of antibodies which has been derived from the maternal supply. These must last until a few weeks after birth, at which point the infant can satisfactorily synthesize its own antibodies. The colostral and subsequent milk antibodies are not absorbed by the infant but act locally in the gut, preventing the growth of undesirable bacteria. These antibodies (secretory IgA in type, see Chapter 8) are not digested and appear in the faeces. Other proteins in milk help to reduce infection. Lactoferrin has a very high affinity for iron and so reduces the amount of free iron available in the gut. Bacteria need free iron to grow and so when the level of free iron in the gut is held low by this protein microbial growth is limited. Non-protein components also act to reduce infection with undesirable bacteria. Milk is rich in lactose, and human milk particularly so. It also contains a factor, possibly a fatty compound, called the 'bifidus factor'. Together, lactose and the bifidus factor lead to the growth of lactose-fermenting bacteria, increasing the acidity of the lower gut and reducing the growth of undesirable bugs. Milk

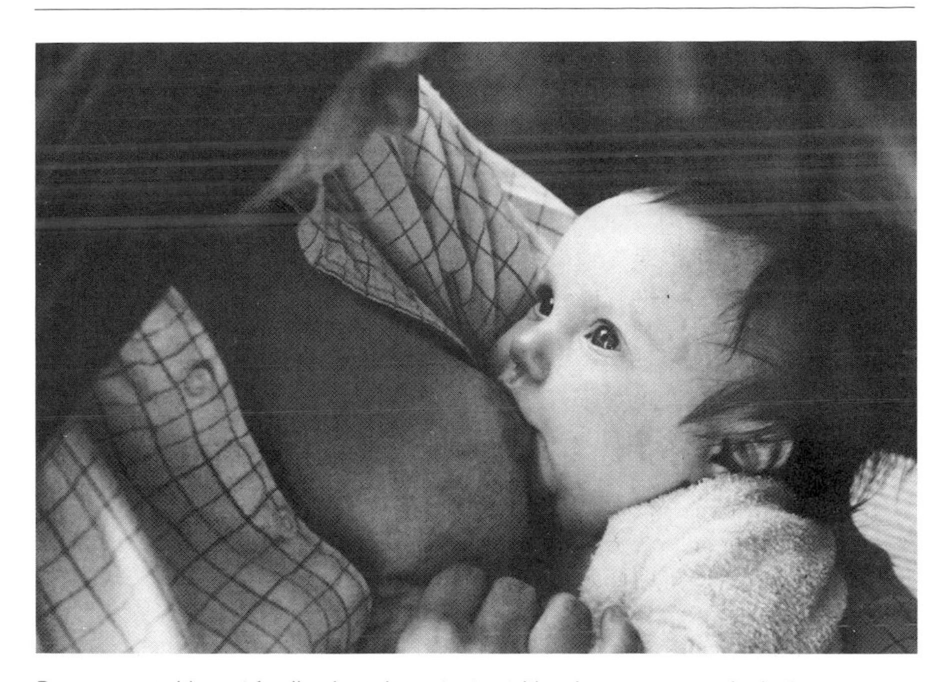

Pregnancy and breast feeding have important nutritional consequences for both the mother and baby and this is discussed further in the text. (Photograph by Nigel Luckhurst.)

also contains cells of the immune system, both lymphocytes which can secrete antibodies and macrophages which can engulf bacteria. All in all, human milk is geared to ensure that the suckling infant is protected from intestinal infection for that very crucial period between birth and the point at which the infant becomes self-sufficient in antibody synthesis.

Cow's milk, on which most milk substitutes are based, does not contain the required antibodies, lactoferrin, the bifidus factor, high levels of lactose or the cells of the immune system. Most modern milk substitutes contain an array of different proteins from cow's milk, are supplemented with minerals and vitamins, and are enriched in polyunsaturated fatty acids. Human milk is a good source of essential fatty acids.

The volume of milk secreted by Western women gradually increases from 560 millilitres per day in the first fortnight after birth to 720 for the next month, and to 750 millilitres per day thereafter for six or seven months. Thereafter, volumes are about 400–600 millilitres per day for the second six months, 300–500 in the second year and, for those brave enough, 200–500 in the third year. Mothers in underdeveloped parts of the world secrete less milk per day although not substantially less. Famine will of course lead to cessation of lactation and to cessation of life.

There has been much controversy over the role of breast feeding in Third World countries and indeed criticism of the multinational companies who market their milk substitutes there. Clearly, breast fed babies receive more protection from infection by means of the constituents of the milk and because the milk as secreted from the breast is relatively sterile. The infant fed on milk substitute loses out on natural immunity and runs the risks of the milk being contaminated during preparation, of its being over- or underdiluted, and of erratic supply depending on household cash flow. An additional benefit of breast feeding in Third World countries is its contraceptive effect. This is often pooh-poohed but merits consideration. It appears that prolactin exerts an antiovulatory effect and a delay in return of menstruation. These effects are dependent on two factors, both of which do not occur to any great extent in developed countries: frequent and intense sucking stimulus (which won't occur with Western-style supplementary feeding) and a low level of maternal fat stores. This was illustrated in a study of the effect of supplementary feeding of Gambian women during lactation. No increase in milk output was observed but by 18 months postpartum, 33% of supplemented women were pregnant, compared to 7% of non-supplemented controls.

Nutrition and athletic performance

Up until 20 years ago, most athletes had worked out for themselves the diets that they felt most comfortable with. With the post-war boom in the craving for athletic success and its confusion with ideological conflicts, the physiology of athletic performance has been studied intensively. It is now generally accepted that athletes require a diet not unlike that advocated for the general public: proteins 10–15% of energy; total fats 30–35%; saturated fats 10%; carbohydrates 50–60%.

Athletes with very high energy demands may need to obtain up to 70% of their energy from carbohydrate. Other than these guidelines, there is

One of the most gruellig athletic events is the triathlon – a combination of running, cycling and swimming that requires considerable stamina. Although diet can be a significant factor in preparing for such events, it is even more important to train properly. (Photographs courtesy of Cambridge Amenities and Recreation Department, organisers of the Cambridge Triathlon. Reproduced by permission of the *Cambridge Evening News*.)

nothing mystical in the diet of athletes. The 'protein' mystique, quite different from the raw steak theory of rugby coaches, may have arisen through the nineteenth-century belief that muscle activity was a protein-consuming exercise. It is not. The main fuel for muscular activity is fat, derived from muscle fat and from fatty acids in blood *en route* from adipose tissue. Training increases the capacity to use fat as a fuel and so conserve for as long as possible the carbohydrate reserves of muscle, i.e. glycogen (see Chapter 2). In a normal 70-kilogram man, there are 12 kilograms of fat reserves or 110 000 kilocalories. In a lean marathon runner of 60 kilograms, there will be but 3 kilograms of fat, representing 23 000 kilocalories, enough incidentally for eight marathons. In contrast, the glucose present in blood would provide a mere 80 kilocalories, and the glucose stored as muscle glycogen, only 1600 kilocalories. However, a 60-kilogram athlete would expend only 2400 kilocalories above maintenance to complete a marathon and hence this 1600 kilocalories of muscle glycogen (i.e. glucose) energy can be vital. Indeed carbohydrate 'loading' to increase energy stores is an integral part of many marathon runners' preparations.* The results of one study, using a bicycle ergometer, are summarized in Table 11.

Clearly, a high-carbohydrate diet can help to sustain athletic effort. Whilst many athletes take special tailor-made vitamin plus mineral supplements, there is little evidence that they exert any beneficial physiological effect.

Nutrition and cancer

Cancer, a major cause of death in the developed world, is a much-feared disease. Cancer patients and their families will go to great lengths to try new cures and consequently are open to exploitation by quacks and charlatans. The truth is that links between diet and cancer are highly tenuous, based on epidemiological data and animal experiments. Nonetheless, there is a growing body of evidence to suggest that cancer is related to obesity, to excess intakes of fats, of proteins and of salt, and to inadequate intakes of fibre and vitamin A. Other nutrients such as selenium, zinc and vitamin E have been associated with cancer, but on somewhat flimsy

* Marathon runners often go on a low-carbohydrate diet for the first few days of the week preceeding the race. This depletes their muscle glycogen reserves. The glycogen-synthesizing enzymes become elevated so as to make the most of what little carbohydrate the body can spare for storage. Just as these enzymes are in top gear, the runner starts three days of a high-carbohydrate diet. Production of muscle glycogen is maximal and the amount of stored glycogen overshoots the normal level. The marathon runner is ready and primed.

Table 11

Diet	Muscle glycogen (% of weight)	Time to exhaustion (minutes)
Low-carbohydrate	0.63	57
High-carbohydrate	3.31	167

evidence. Indeed the subject of nutrition and cancer is characterized by its low content of hard data and high content of conjecture. One reviewer has summed it up thus:

The role of nutrition and diet in the etiology of cancer is in need of continuing research of all kinds – epidemological, biochemical and nutritional. Unfortunately, research in this area is being delayed and hindered by premature pronouncements by politicians, less than knowledgeable scientists, and sensationalists looking for notoriety to help sell an article, a magazine or a book. These reports are usually based on unconfirmed hypotheses and incomplete data and on reports not given a critical peer review.

Dental caries and sucrose

Common table sugar, sucrose, is comprised of two molecules, glucose and fructose. It is strongly linked with dental caries. Certain bacteria which inhabit the mouth convert sugar into various acids which eat away tooth enamel, allowing a mesh or plaque to grow over the rotting tooth and thus prevent access by protective factors in saliva. In this respect, all sugars are equally culpable, even brown sugar and honey, and while the absolute intake is important, the frequency of intake can compound matters. A regular intake of sugar-rich sweets throughout the day is the perfect means of ensuring that the acid- and plaque-forming bugs operate at maximal capacity in tooth destruction. Dental caries is a preventable disease and the resources currently wasted in fillings and extractions could be better spent elsewhere. For example, in the UK, where fluoridation of water is commonplace, some 30 million teeth are filled and 5 million teeth extracted each year at an annual cost of £400 million. This is more than the health costs of all cancers put together.

Novel proteins

If the 36 billion people on Earth each ate 70 grams of protein per day, the annual global requirement for protein would be 92 megatonnes (1 megatonne = 1 million tonnes; 1 tonne = 1000 kilograms). At first sight this requirement should be readily met by the annual production of protein: 100 megatonnes of cereal protein and 35 megatonnes of oil-seed protein. However, little of this is directly available for human consumption. Most of it, almost 95% in North America, is fed to livestock for meat and milk production. The possibility that this state of affairs could precipitate a protein short-fall has led scientists to think about novel proteins. The most successful of these have been the textured vegetable proteins (TVP) based on soya beans. Indeed many foods listing 'vegetable proteins' as an additive actually contain TVP. Nutritionally speaking, there is nothing wrong with these proteins. In fact many would see them as advantageous, allowing protein requirements to be met while keeping the fat content low. The only disadvantage of TVP is their moderate levels of soluble but indigestible carbohydrates which can be rapidly fermented by bacteria in the large intestine. The result is flatulence, discussed previously. Another potential source of novel proteins is microbes. Indeed, there was considerable investment in this technology by the petrochemical industry in the 1960s. They grew bacteria on waste products of the oil industry but their efforts were thwarted by the energy crisis and soaring oil prices of the early 1970s. Such microbial protein would pose problems which don't occur with TVP. Each microbe has a nucleus containing its genetic code, contained in the DNA, and this code is translated in protein synthesis by the similar chemical RNA. Microbial protein contains quite high levels of these 'nucleic acids' and man cannot take much without risking gout! The nucleic acids are metabolized to a compound called uric acid which is only poorly soluble and crystallizes out in the joints. The old idea that port and gout went together was probably because port was kept in lead containers and the 'gout' was a form of lead poisoning. Recently, a British company launched a new protein source based on fungal protein. It has a low nucleic acid content and can grow on carbohydrate slurries. It also has the advantage of being filamentous by nature, thus lending itself to be spun into fibres for simulated steaks. None of these is likely to compete seriously with conventional protein in our diet but the technology may one day be useful.

Nutrition and alcohol

In vino veritas:
In beer which is lower class,
The degree of veracity
Depends on capacity.

Composed by a Polish colleague, this little rhyme hides the truth. Perhaps alcohol does make tongues wag, but in the sober light of day no one, not least the heavy drinker, admits his or her capacity. It is therefore difficult to estimate alcohol intake and its contribution to caloric intake. The general figure put forward is 5% of kilocalories. Although alcohol has more calories per gram than carbohydrate, it is no more fattening, probably because it raises basal metabolic rate. Moderate intakes of alcohol are believed to reduce the risk of coronary heart disease by increasing the levels of high-density lipoprotein, which takes cholesterol from the tissues to the liver for excretion. Even moderately excessive intakes of alcohol can lead to high levels of another lipoprotein class, the very low-density lipoproteins. No one is quite sure whether this increases the risk of heart attack. The big danger of excess alcohol intake is liver damage, particularly among alcoholics who derive up to half their caloric intake from alcohol. Because alcohol has little or no nutrients, alcoholics become deficient in many minerals and vitamins. This is compounded by damage to the gut which reduces absorption and it is these malnourished souls who are at most risk of liver disease.

Vegetarianism

The domestication of animals was a milestone in man's evolution. Why therefore did many decide against the killing of animals for meat? How did vegetarianism begin? Reay Tannahill in her magnificent work *Food in History* gives an excellent summary.

In about 2000 BC the Aryans arrived in India and brought with them their high-yielding cattle. In time these were replaced by native cattle of a

Religion has always had a significant influence on the type of diet adopted by followers. The origins of vegetarianism seem to be tied in with the creeds of Buddha – though today the vegetarian diet is widely adopted for ethical reasons alone. (Reproduced with permission: COMPIX – Commonwealth Institute.)

tropical breed which had lower milk yields. As the Aryans spread about India they needed to guard their cattle. At first any meat but fertile-cow meat was permitted. Then, no meat was to be consumed but animals were still kept for ritual slaughter for the priests. Such was the demand for cattle for ritual slaughter by the priests that peasants were being hard done by. Two new sects emerged to oppose the caste divisions of the Aryans, their violence and their penchant for ritual slaughter. The founders were the Buddha and Mahavira. Vegetarianism was established and not for any sentimental love of animals. Today many people have adopted vegetarianism because of a sentimental love of animals or more correctly a distaste of their exploitation. Life as a vegetarian can have its social difficulties. It has no nutritional difficulties. Indeed, it is in light of present expert opinion, a very healthy diet: high-fibre, low-fat and low-sodium.

13
Health and well-being

In North America, the term used to describe the diet set forth by experts in nutrition is 'prudent', hence the 'Prudent Diet'. Now prudence is but one of the seven cardinal virtues. It is a moral term. The fact that it defines the diet which many nutritional experts throughout the Western world see as presently desirable, tells us much about their quest for nutritional reformation. They see heart disease, cancer, diabetes, hypertension, diverticulitis and obesity, the major causes of death and disease in adulthood, as the wages of sin – consequences of our rich diet and slothful lifestyle. By inference, it is a lifestyle lacking in virtue. It is imprudent. Thus if we change our ways, we can restore our virtue. We can eat prudently, live better, healthier, longer lives. We are asked to accept their views, their consensus of informed opinion. In essence, their advice can be summarized into six edicts:

1 Identify your ideal weight for height, achieve it and maintain it.
2 We should reduce our intake of fat from its present level of 40% + of caloric intake to between 30% and 35%.
3 In so reducing our fat intake we should seek to reduce our intake of

saturated fats and, if needs be, increase our intake of polyunsaturated fats.

4 We should halve our intake of dietary cholesterol.
5 We should reduce our salt intake from about 15 grams per day to between 5 and 10 grams per day.
6 Sugar should not provide more than 10% of caloric intake.

The general public and their educators, the media, general medical practitioners, dietitians, health visitors and so forth, are likely to adopt these views because 'they' are the experts and 'they' should know what they're talking about. Many scientists reject this advice, because 'they' have not obtained proof in the scientific sense. Consequently, they can be no more certain of justifying their expectation than a politician can of some new fiscal policy generating a given number of new jobs. That is not the way of science. Normally a scientist would conduct an experiment which would constitute reproducible evidence that a certain expectation is justified. The diet–disease link does not lend itself to such scientific methodology. It cannot be studied fully within the confines of the laboratory. It is open research. Being multifactorial, the link between diet and disease is particularly difficult to study. Which factor do you control: smoking, blood pressure, blood cholesterol, diet, body weight? Some of them? All of them? And who do you study? The normal majority or the abnormal minority? In spite of these and many other limitations, many billions of dollars have been spent to answer the question: can we reduce the incidence of disease by mass alteration of lifestyle and diet? In general, the results have been inconclusive. Therefore many scientists, particularly those working at the molecular and cellular forefront of diet–disease issues, are sceptical about the expected end-points or benefits of the so-called Prudent Diet. They are paid to be sceptical. That is the social function of science and its practitioners. Their scepticism will temper the dogmatic crusade for a nutritional reformation. But they will not succeed, for too many people in too many places have staked their careers and reputations on just such a change, and the signs are they will have their way. Their Prudent Diet will probably not be harmful and it might indeed do good. But could they have put the cart before the horse?

There is a third option which inverts the problem for its solution. Health is not merely the absence of disease. It is the 'attainment and maintenance of the highest state of mental and bodily vigour of which any individual is capable'. It implies vivacity (*vivere* = to live; *actias* = abounding in). It is the antithesis of a slob-like existence. It implies an appetite for life in all its aspects, of physical activity, of literature and the arts, of conversation and philosophy, of comradeship, of fun and laughter, of the capacity to love and to grieve. From the pursuit of such an outlook on life will emerge an outlook on food, which sees diet, not as a physiological fuel, not as a prescriptive medicine to stave off disease but as both fun and functional. It implies variety of cuisines, of menus and of meal patterns. It is to be gastronomically pleasing and both mentally and physically invigorating. And if, as a result of such vivacity, our diet approaches that proposed by the expert committees, then well and good. But to approach the problem by seeing their Prudent Diet as a response to our 'sinful' diet and to adopt their diet for medical reasons is to begin at the wrong end. Food is not medicine. It is, for the lucky ones, fun.

Dr William Carlyon, Director of Health Education with the American Medical Association, uses the term wellness: '. . . a condition of wholeness, happiness, high-quality living with dignity, purpose and meaning, alive clear to your fingertips, tingling with vitality'. He points out that trying to achieve wellness goals with disease prevention concepts is like using a map of China to explore Africa.

Some dietary modification may reduce the incidence or alter the pattern of the so-called diseases of civilization, but it will not create health. Each of us, within our own social, economic and domestic constraints must find our own way to wellness and, collectively, society must provide the opportunity for all of us to do so. We cannot eat our way to wellness.

This book ends with a poem by Dr Carlyon, 'The Healthiest Couple':

They brush and they floss
with care every day,
But not before breakfast
of both curds and whey.

He jogs for his heart
she bikes for her nerves;
They assert themselves daily
with appropriate verve.

He is loving and tender
and caring and kind,
Not one chauvinist thought
is allowed in his mind.

They are slim and attractive
well dressed and just fun.
They are strong and well-immunized
against everything under the sun.

They are sparkling and lively
and having a ball.
Their diet? High fiber
and low cholesterol.

Cocktails are avoided
in favour of juice;
Cigarettes are shunned
as one would the noose.

They drive their car safely
with belts well in place;
at home not one hazard
ever will they face.

1.2 children they raise,
both sharing the job.
One is named Betty,
.2 is named Bob.

And when at the age of
two hundred and three
they jog from this life
to one still more free,

They'll pass through those portals
to claim their reward
and St. Peter will stop them
"just for a word".

"What Ho" he will say
"You cannot go in.
This place is reserved
For those without sin".

"But we've followed the rules"
she'll say with a fright
"We're healthy" –
"Near perfect" –
"And incredibly bright".

"But that's it" will say Peter,
drawing himself tall.
"You've missed the point of living
By thinking so small".

"Life is more than health habits,
Though useful they be
It is purpose and meaning,
and grand mystery".

"You've discovered a part
of what makes humans whole
and mistaken that part
for the shape of the soul".

"You are fitter than fiddles
and sound as a bell,
Self-righteous, intolerant
and boring as hell".

Appendix 1

The list of foods in Table A1 is not intended to be complete and comprehensive but to serve as a guide to the nutrient content of the food groups. Only those nutrients which are present under focus by health educationalists are included. Therefore, the list excludes all the minerals, trace elements and vitamins. The purpose of including this list is to help you to be more familiar with the nutrient content of foods and how these are influenced by cooking and processing.

A quick glance down the fat column will tell you what foods to select to lower fat intake, namely vegetables, fruit, cereals, legumes and white fish. You will also notice that these food categories are the richest source of fibre and in general, are lowest in sodium, hence my previous reference to the greengrocer as your local health food store. If like most people in Western Society you want to eat meat, you will see that grilling rather than frying will make it easier to moderate fat intake. If you compare boiled potatoes with either roast or chipped potatoes you can again see how cooking methods influence fat intake. Fresh foods, in general, tend to be lower in sodium than processed food. Thus, while raw bran contains 28 milligrams sodium per 100 grams, the value for all bran is 1670! Similarly, fresh peas have a trace of sodium while canned peas have 330 milligrams per 100 grams. You will also see that sweets, cakes, pastries, sausages and peanuts are quite high in fat, often referred to as 'invisible fat'. Sugars refers to all 'simple' sugars including lactose, sucrose and glucose (see Chapter 2). Sucrose or 'table sugar' is found in high concentrations in fancy cakes and biscuits, soft drinks and jams. You will also see that sugar is added to many foods as a flavouring agent, e.g. muesli and baked beans.

Table A1 *The composition of selected foods (per 100 grams)*

Nutrient	Energy	Protein	Fat	Total carbo-hydrate	Sugars	Sodium	Fibre	Typical serving weight
Units	(kcal)	(g)	(g)	(g)	(g)	(mg)	(g)	(g)
White flour	339	9.3	1.2	77.5	1.4	350	3.7	–
White bread	233	7.8	1.7	49.7	2.7	540	1.4	30
Wholemeal bread	216	8.5	2.7	41.8	2.1	540	8.5	40
Cooked porridge	44	1.4	0.9	8.2	tr	300	0.8	200
Cornflakes	368	8.6	1.6	85.1	7.4	1160	11.0	30
All-bran	273	15.1	5.7	43.0	15.4	1670	26.7	30
Muesli	368	12.9	7.5	66.2	26.2	180	7.4	30
Rice, boiled	123	2.2	0.3	29.2	tr	2	0.8	250
Chocolate biscuit	524	5.7	27.6	67.4	43.4	160	3.1	20
Plain biscuit	471	9.8	20.5	66.0	16.4	440	5.5	15
Fancy iced cake	407	3.8	14.9	68.8	54.0	250	2.4	60
Plain fruit cake	354	5.1	12.9	57.9	43.1	250	2.8	60
Doughnuts	349	6.0	15.8	48.8	15.0	60	–	60
Whole milk (per 100 ml)	65	3.3	3.8	4.7	4.7	50	–	150ml
Processed cheese	311	21.5	25.0	tr	tr	1360		25
Cheddar cheese	406	26.0	33.5	tr	tr	610	–	25
Trout, baked	135	23.5	4.5	–	–	88	–	–
Egg, boiled	147	12.3	10.9	tr	tr	140	–	55
Egg, fried	232	14.1	19.5	tr	tr	220	–	60
Pizza	234	9.4	11.5	24.8	–	340	–	150
Quiche Lorraine	391	14.7	28.1	21.1	–	610	–	150
Butter (mini pack)	740	0.4	82.0	tr	tr	870	–	10
Margarine	730	0.1	81.0	tr	tr	800	–	10
Rashers, fried	477	24.1	42.3	tr	tr	1870	–	2=40
Rashers, grilled	416	24.9	35.1	tr	tr	2000	–	2=40
Rump steak, fried	246	28.6	14.6	tr	tr	54	–	150
Rump steak, grilled	218	27.3	12.1	tr	tr	55	–	150
Mince, stewed	229	23.1	15.2	tr	tr	320	–	120

Table A1 (*cont.*)

Nutrient	Energy	Protein	Fat	Total carbo-hydrate	Sugars	Sodium	Fibre	Typical serving weight
Units	(kcal)	(g)	(g)	(g)	(g)	(mg)	(g)	(g)
Lamb chops, grilled	355	23.5	29.0	tr	tr	72	–	120
Pork chops, grilled	332	28.5	24.2	tr	tr	84	–	120
Chicken, roast	216	22.6	14.0	tr	tr	72	–	100
Liver, fried	254	26.9	13.2	7.3	–	170	–	60
Black pudding, fried	305	12.9	21.9	15.0	–	1210	–	60
Pork sausages, fried	317	13.8	24.5	11.1	–	1050	–	2 = 120
Pork sausages, grilled	318	13.3	24.6	11.5	–	1000	–	2 = 120
Corn beef, tinned	217	26.9	12.1	0	–	950	–	60
Beefburgers, fried	264	20.4	17.3	7.0	–	880	–	90
Cornish pasty	332	8.0	20.4	31.1	–	590	–	160
Sausage roll	463	8.1	31.8	38.4	–	580	–	100
Cod, grilled	95	20.8	1.3	–	–	91	–	150
Fish fingers, fried	233	13.5	12.7	17.2	–	350	–	5 = 100
Beans, canned, baked	64	5.1	0.5	10.3	5.2	480	7.3	125
Runner beans, boiled	19	1.9	0.2	2.7	0.8	1	3.2	100
Brussel sprouts, boiled	18	2.8	tr	1.7	1.6	2	2.9	7 = 70
Cabbage, boiled	9	1.3	tr	1.1	1.1	8	2.5	100
Cauliflowers, boiled	9	1.6	tr	0.8	0.8	4	1.8	100
Lentils, boiled	99	7.6	0.5	17.0	0.8	12	3.7	100
Fresh peas, boiled	52	5.0	0.4	7.7	1.8	tr	5.2	60
Canned peas, boiled	80	6.2	0.4	13.7	1.3	330	7.9	60
Carrots, boiled	19	0.7	tr	4.3	4.2	50	3.1	100
Green peppers, raw	15	0.9	0.4	2.2	2.2	2	0.9	$\frac{1}{4}$ = 15
Potatoes, boiled	80	1.4	0.1	19.7	0.4	3	1.0	250
Potatoes, mashed	119	1.5	5.0	18.0	0.6	24	0.9	250
Potatoes, baked	105	2.6	0.1	25.0	0.6	8	2.5	250
Potatoes, roast	157	2.8	4.8	27.3	–	9	?	250
Potatoes, fried	253	3.8	10.9	37.3	–	12	?	12 = 200
Tomatoes	14	0.9	tr	2.8	2.8	3	1.5	120
Lettuce	12	1.0	0.4	1.2	1.2	9	1.5	2 = 10 (leaves)
Mushrooms, raw	13	1.8	0.6	0	0	9	2.5	3 = 30
Onions, raw	23	0.9	tr	5.2	5.2	10	1.3	60
Turnips, boiled	14	0.7	0.3	2.3	2.3	28	2.2	100
Apples	46	0.3	tr	11.9	11.8	2	2.0	120
Avocadoes	223	4.2	22.2	1.8	1.8	2	2.0	$\frac{1}{5}$ = 150
Bananas	79	1.1	0.3	19.2	16.2	1	3.4	150
Grapes	63	0.6	tr	16.1	16.1	2	0.9	20 = 100
Grapefruit (whole)	22	0.6	tr	5.3	5.3	1	0.6	200
Lemons	15	0.8	tr	3.2	3.2	6	5.2	2 = 15 slices
Watermelon	21	0.4	tr	5.3	5.3	4	–	260
Peaches	37	0.6	tr	9.1	3.1	3	1.4	100

Table A1 (*cont.*)

Nutrient	Energy	Protein	Fat	Total carbo-hydrate	Sugars	Sodium	Fibre	Typical serving weight
Units	(kcal)	(g)	(g)	(g)	(g)	(mg)	(g)	(g)
Pears	29	0.2	tr	7.6	7.6	1	1.7	120
Pineapple	46	0.5	tr	11.6	11.6	2	1.2	80
Plums	38	0.6	tr	9.6	9.6	2	2.1	3 = 100
Strawberries	26	0.6	tr	6.2	6.2	2	2.2	10 = 100
Almonds	565	16.9	53.5	4.3	4.3	6	14.3	10 = 15
Brazil nuts	619	12.0	61.5	4.1	1.7	2	9.0	5 = 20
Roasted peanuts	570	24.3	49.0	8.6	3.1	440	8.1	30 = 25
Sugar, white	394	tr	0	105	105	tr	0	1 = 5 (tspn)
Sugar, brown	394	tr	0	105	105	6	0	1 = 5 (tspn)
Jam	261	0.6	0	69	69	16	1.1	10
Boiled sweets	327	tr	tr	87	87	25	0	1 = 5
Mixed toffee	430	2.1	18.2	71	70	320	–	1 = 5
Oxtail soup	44	2.4	1.7	5.1	0.9	440	–	200

(1) To convert milligrams of sodium to grams of salt, you multiply by 2.5 and divide by 1000.
(2) To convert intakes of protein, carbohydrate and fat from absolute weights to percentage energy, you first multiply the weights by the appropriate factors (4.0 for carbohydrate, 4.0 for protein and 9.0 for fat) to obtain the caloric value. Now add these together and express each caloric value as a % of total.
(3) Based on *The Composition of Foods* (4th edition). A.A. Paul & D.A.T. Southgate. HMSO, London, 1979.
(4) The values for average servings refer to edible matter.

Evaluation of Dietary Data

Case

These data are from a middle-aged housewife. You can almost hear her proudly proclaim that she is quite conscious of her diet: 'I don't touch salt, ever, or sugar for that matter. I'm very conscious of dietary fibre and I always use a polyunsaturated margarine. I insist on lean meat. However, I'm not a fanatic and I do try to enjoy my food without overdoing the health bit'.

Table A2. *A 24-hour sample of the diet of a middle-aged housewife*

Meal	Food	Estimated weight
Breakfast	All-Bran	30 g
	Milk ½ cup	75 ml
	2 slices of wholemeal bread	80 g
	Margarine	20 g
	Marmalade	20 g
	Milk for coffee (2 cups)	80 ml
Lunch	Fried steak	150 g
	Chipped potatoes	150 g
	Boiled carrots	100 g
	Boiled peas	60 g
	Fancy iced cake	60 g
	Milk for coffee (2 cups)	80 ml
Dinner	Middle bacon, fried (2 slices)	40 g
	4 Fried pork sausages	120 g
	2 slices wholemeal bread	80 g
	Margarine	20 g
	Apple	100 g
	Milk for coffee (2 cups)	80 ml
Supper	Fancy iced cake	60 g
	Milk for coffee (2 cups)	80 ml

When the data in Table A2 are analysed they reveal the following:

Daily energy intake	(Kcal)	2924
Daily fibre intake	(g)	40
Daily salt intake	(g)	11
Daily sugar intake	(g)	65
% Energy from		
Fat		51
Carbohydrate		24
Sucrose		9
Protein		16

If the nutritional targets described in Appendix 2 are used as our yardstick, this lady scores poorly. In spite of her righteous claims, she consumes, unwittingly, quite a good deal of salt and sugar. Her fibre intake is very satisfactory but one would be seriously critical of both her total energy and that fraction of it from fat. Almost all of her sugar intake comes from fancy iced cakes which also contribute 8% of her fat intake. Not surprisingly, meat and meat products supply 41% of her fat intake with chips providing a further 17%. Equally, meats contribute 48% of her salt intake while cereals contribute a further 31%.

How should she set about improving matters? To begin with, we could advise her against chips and coax her into eating more baked and boiled potatoes. Indeed she would ideally need to increase her overall carbohydrate intake so perhaps additional wholemeal bread would be advised without additional margarine. In place of fried steak, we could recommend roast chicken and in place of bacon and sausage we might advise grilled cod. Apples could substitute for her fancy iced cake and let us add a tomato to her evening meal. The modified diet is shown in Table A3, and the nutrient analysis is as follows:

Daily energy intake	(Kcal)	2095
Daily fibre intake	(g)	48
Daily salt intake	(g)	7
Daily sugar intake	(g)	negligible
% Energy from		
Fat		34
Carbohydrate		46
Sucrose		negligible
Protein		20

Now, with appropriate remedial action, her nutrient intakes are deemed satisfactory if not excellent.

Table A3. *A modified diet for the middle-aged housewife*

Meal	Food	Estimated weight
Breakfast	All-Bran	30 g
	Milk $\frac{1}{2}$ cup	75 ml
	2 slices of wholemeal bread	80 g
	Margarine	13 g
	Marmalade	20 g
	Milk for coffee (2 cups)	80 ml

Table A3 (*cont.*)

Meal	Food	Estimated weight
Lunch	Roast Chicken	150 g
	Baked potatoes	150 g
	Boiled carrots	100 g
	Boiled peas	60 g
	Apple	100 g
	Milk for coffee (2 cups)	80 ml
Dinner	Grilled Cod	100 g
	Boiled potatoes	150 g
	Tomato	60 g
	Wholemeal bread (2 slices)	80 g
	Margarine	13 g
	Apple	100 g
	Milk for coffee (2 cups)	80 ml
Supper	2 slices of wholemeal bread	80 g
	Margarine	13 g
	Apple	100 g
	Milk for coffee (2 cups)	80 ml

Case 2

These data are from a young man who has to some extent embraced the quest for nutritional reformation as part of a healthy lifestyle which includes both exercise (jogging, cycling, swimming) and relaxation (music and reading). He chooses semi-skimmed milk but rather than take margarine, which he simply does not like, he eats butter, sparingly he feels. He has been told of the goodness in nuts and that fish is an excellent food. He will not eat fries as a rule and genuinely likes porridge. All his bread is wholemeal. Like the previous case, he avoids all sugar and salt. The nutritional composition of his diet is as follows (see Table A4).

Daily energy intake	(Kcal)	2181
Daily fibre intake	(g)	26
Daily salt intake	(g)	10
Daily sugar intake	(g)	negligible
% Energy from		
Fat		32
Carbohydrate		39
Sucrose		negligible
Protein		18
Alcohol		11
Excluding alcohol:		
Fat		36
Carbohydrate		44
Protein		20

This young man is somewhat confused about his alcohol intake. According to the NACNE report (see Appendix 2) he is consuming twice the recommended level. However, according to the COMA report (Appendix 2) his alcohol intake of 36 millilitres is well below the upper limit of 100. So he intends to believe COMA on this occasion. When it is pointed out to him that his fat intake (expressed as a percentage of non-alcohol energy) is slighly over the top, he is not impressed. His diet is pretty good, he feels.

Table A4. *A 24-hour sample of the diet of a young man*

Meal	Food	Estimated weight
Breakfast	Porridge	300 g
	Semi-skimmed milk	60 ml
	Wholemeal bread (1 slice)	40 g
	Butter	10 g
	Tea with semi-skimmed milk (2 cups)	80 ml

Table A4 (*cont.*)

Meal	Food	Estimated weight
Lunch	2 cartons yogurt	400 g
	1 apple	100 g
	Almonds	15 g
	Tea with semi-skimmed milk (2 cups)	80 ml
Dinner	Baked trout	150 g
	Boiled peas	60 g
	Boiled potatoes	200 g
	Wholemeal bread (2 slices)	80 g
	Butter	20 g
	Tea with semi-skimmed milk (2 cups)	80 ml
Supper	Cheddar cheese	25 g
	Wholemeal bread (1 slice)	40 g
	Butter	10 g
	Red Wine	350 ml

Conclusions

The diet of Case 1 broke most of the rules, but with modification it improved drastically. What we must ask is how realistic this modification was. If the lady in question had time to listen and was sufficiently educated and receptive, we could persuade her that her diet was poor by the standards set out. We would have more difficulty with a busy, working mother with more on her mind than diet. Either way, we would simply have taken the horse to water so to speak. Whether it chooses to drink is another matter. Realistically speaking, this lady's diet requires a major overhaul and unless she is motivated and has the time and resources, she will make only some modifications, perhaps grilling her meat and cutting down on chips. But if she, her husband or her children are unemployed or uncertain of their job future, or if the family suffer the stress of inner urban decay or marital or financial problems or illness, why should she bother changing her diet because some well-heeled middle class do-gooder offers her and her family the somewhat uncertain *chance* of a reduction in the *risk* of disease. Under these circumstances a modified diet will do little to promote health or well-being. Nonetheless, if the public health experts deem their goals as socially desirable, every citizen has a right to be told of them.

In the second case, this young man has made a conscious effort to alter his lifestyle including diet to one nearer the concept of health and wellness. Further modifications are not negotiable. He is keen to point out that within his own specifications of eating wholegrain cereals, low-fat dairy products, of avoiding salt and sugar, he intends to eat as he pleases and to pot with the consequences. He wants to enjoy life.

Anyone who feels that the mere issuing of dietary guidelines will alter a nation's diet is naive. Anyone who thinks it is easy to achieve the generally recommended goals should try it themselves. It is not easy in every day life. We are not rats and our diet does not come pelleted in paper sacks. Not yet at least!

Appendix 2

Summary recommendations of expert committees for the general population.

*Committee on medical aspects of food policy (COMA report)**

Fat	35% of energy
Saturated fat	15% of energy
Polyunsaturated fat	3.5–6.8% of energy
Cholesterol	No specific recommendation
Sugar	Intake should not be increased further
Alcohol	Avoid excess, be moderate. Excess for men is defined as 100 millilitres/day and for women it is 64 millilitres/day
Salt	Intake should not increase further and consideration should be given to ways and means of decreasing it
Fibre	There are no specific recommendations about the intake of fibre. However, advantages are seen in using fibre-rich carbohydrates to compensate for a reduced fat intake
Body weight	Obesity should be avoided by a combination of appropriate food intake and exercise

**Appointed by the Department of Health & Social Security, England.*

*National Advisory Committee on Nutrition Education (NACNE Committee)**

Targets for the short-term (15-year programme)	
Fat	35% of total energy
Saturated fat	15% of total energy
Polyunsaturated fat	5% of total energy
Sugar	12% of dietary energy
Fibre	Increase to 25 grams/day
Salt	Reduce by 10%, i.e. not to exceed 9 grams/day
Alcohol	Moderate to 5% of energy
Body weight	Maintain optimal body weight and exercise regularly

*Ad-Hoc Committee

Joint Committee of the United States Department of Health & Human Services, and Department of Agriculture

1. Eat a variety of foods
2. Maintain reasonable weight
3. Avoid too much fat, saturated fat and cholesterol
4. Eat foods with adequate starch and fibre
5. Avoid too much sugar
6. Avoid too much sodium
7. Drink in moderation

American Heart Association – Nutrition Committee

Fat	30% of energy
Saturated fat	15% of energy
Polyunsaturated fat	Increase intake Do not exceed 10% energy
Carbohydrates	Increase intake of complex carbohydrates to 45–55% of energy
Cholesterol	Reduce to 300 milligrams/day
Body weight	Achieve ideal weight through diet exercise

Further Reading

T. G. Taylor (1978). *Principles of Human Nutrition: Institute of Biology, Studies in Biology, No. 94.* Edward Arnold, London.

T. G. Taylor (1982). *Nutrition and Health: Institute of Biology, Studies in Biology.* Edward Arnold, London.

A. E. Bender (1975). *The Facts of Food.* Oxford University Press.

Nutrition Reviews: Present Knowledge in Nutrition. 5th Edition (1984). Nutrition Foundation Inc., 888 Seventeenth Street, NW, Washington DC 20006, USA.

Reay Tannahill (1975). *Food in History.* Paladin, St. Albans, Herts.

R. Barker (1968). *Understanding the Chemistry of the Cell: Institute of Biology, Studies in Biology, No. 13.* Edward Arnold, London.

G. C. Birch & K. J. Parker (eds) (1983) *Dietary Fibre.* Applied Science Publishers, London.

G. E. Inglett & S. I. Falkehag (eds) (1979). *Dietary Fibers: Chemistry and Nutrition.* Academic Press, New York.

Stock & Rothwell (1982). *Obesity and Leanness: Basic Aspects.* John Libbey, London.

J. S. Garrow (1981). *Treat Obesity Seriously: A Clinical Manual.* Churchill Livingstone, London.

B. Feldman (ed.) (1983). *Nutrition and Heart Disease.* Churchill Livingstone, New York.

J. Vitale (1976). *Vitamins: A Scope Publication.* Upjohn Company, Kalamazoo, Michigan.

M. H. Lessof (ed.) (1984). *Allergy: Immunological and Clinical Aspects.* John Wiley & Sons, London.

Herbert (1981). *Nutrition Cultism: Facts and Fictions.* George F. Stickley Co., Philadelphia.

V. Herbert & S. Barrett (1982). *Vitamins and Health Foods: The Great American Hustle.* George F. Stickley Co., Philadelphia.

Index